Cambridge IGCSE™

Chemistry

Practical Skills

Bryan Earl
Doug Wilford

We have carried out a health and safety check of this text and have attempted to identify all recognised hazards and suggest appropriate cautions. However, the Publishers and the authors accept no legal responsibility on any issue arising from this check; whilst every effort has been made to carefully check the instructions for practical work described in this book, it is still the duty and legal obligation of schools to carry out their own risk assessments for each practical in accordance with local health and safety requirements.

For further health and safety information (e.g. Hazcards) please refer to CLEAPSS at www.cleapss.org.uk

Photo credits: p.8 © Martyn F Chillmaid; p.9 © burnell1 - Fotolia

Orders: please contact Hachette UK Distribution, Hely Hutchinson Centre, Milton Road, Didcot, Oxfordshire, OX11 7HH. Telephone: +44 (0)1235 827827. Email education@hachette.co.uk Lines are open from 9 a.m. to 5 p.m., Monday to Friday. You can also order through our website: www.hoddereducation.com

ISBN: 978 1 3983 1051 3

This edition published in 2021 by
Hodder Education,
An Hachette UK Company
Carmelite House
50 Victoria Embankment
London EC4Y 0DZ

www.hoddereducation.com

Impression number 10 9 8 7 6 5 4 3 2 1

Year 2024 2023 2022 2021

Cover © Björn Wylezich - stock.adobe.com

Typeset in India by Aptara Inc.

Printed and bound by CPI Group (UK) Ltd, Croydon, CR0 4YY

A catalogue record for this title is available from the British Library.

Contents

How to use this book

This Practical Skills Workbook will help you keep a record of the practicals you have completed, as well as your results and conclusions. It covers the practical parts of Cambridge IGCSE and IGCSE (9-1) Chemistry syllabuses (0620/0971) for examination from 2023. This resource provides additional practice for the practical skills required by the syllabus with a focus on the investigation-focused learning objectives.

This resource covers Core and Supplement content. Supplement investigations and questions are indicated by a lined box around the text, as shown below.

9 Explain why the white solid forms closer to this end of the tube.

....

Some practicals also have Going Further sections, which provide additional questions that apply the scientific theory learned from the practical to different contexts. These go beyond Core and Supplement level and can be used as stretch activities.

GOING FURTHER

How do you think the method you used today to electrolyse brine could be upgraded to a large scale?

....

Completing the investigations

At the start of each investigation, we have provided a brief piece of context to help explain how the science behind the practical ties into the wider syllabus. Key terms and equations that are relevant to each investigation are also provided.

The aim of each practical is then laid out, along with a list of apparatus needed to complete the practical as suggested. Your teacher will inform you whether they have decided to change any of the equipment and if the method needs to be adapted as a result.

Before you begin the practical and start on the method it is vital that you read and understand the safety guidance, as well as taking any necessary precautions. Once you have carried out a risk assessment and made everything safe, you should check with your teacher that it is appropriate to begin working through the method.

The method itself is presented in a step-by-step fashion and you should read it through at least once before starting, making sure you understand everything. Then, ensuring that you don't miss anything out, you should work through the practical safely. Tips may be provided to help with particularly problematic steps.

Questions and answers

Within each practical there are clear sections laid out for observations where you should record your results as you complete the practicals. Scaffolded questions are also provided to help you develop conclusions and evaluate the success of the experiment.

At the end of the book are past paper questions that relate to the practicals within this book and provide useful exam practice. Your teachers may decide to set this as part of the lesson, or at a later date.

Answers are provided in the accompanying *Cambridge IGCSE™ Chemistry Teacher's Guide with Boost Subscription* (ISBN 9781398310520) at boost-learning.com.

Experimental skills and abilities

Skills for scientific enquiry

The aim of this book is to help you develop the skills and abilities needed to perform practical laboratory work in chemistry. Before you start any practical work, you need to make sure you are aware of the paramount importance of working **safely** and so this is covered first. Then we introduce the apparatus and measuring techniques that you will use most often.

This is followed by a section on how to make and record measurements accurately. Methods for handling the observations and data you have collected will then be described.

Finally, we discuss how to plan, carry out and evaluate an investigation. You should then be ready to work successfully through the experiments and laboratory activities that follow.

Safety

In all your practical exercises and investigations, materials will be used which, although familiar in many cases, are of a potentially hazardous nature, and appropriate care and precautions should be taken. If in doubt, ask your teacher to make the final decision depending on the circumstances at the time. Also, in certain circumstances disposable gloves and fume cupboards will be required. Eye protection should be worn at all times.

Here are a few simple precautions to help ensure your safety when carrying out experiments in the laboratory.

- **Eye protection MUST be worn for any practical involving chemicals.**
- **Always wear stout or sensible shoes (not plimsolls or trainers!)**– to protect your feet if a heavy weight should fall on them.
- **Hot liquids and solids** – set in a safe position where they will not be accidentally knocked over; handle with caution to avoid burns.
- **Toxic materials** – materials such as mercury are toxic; take care not to allow a mercury thermometer to roll onto the floor and break. Other toxic substances such as bromine, chlorine and lead may be present and appropriate information will be given to you on the worksheet and/or by your teacher.
- **Tie back long hair** – to prevent it being caught in a flame.
- **Personal belongings** – leave in a sensible place so that no one will trip over them!
- **Protect eyes and skin from contact with corrosive and harmful chemicals** – any reagent used for any of the experiments in this book must be treated with caution. **Ask for your teacher's advice before handling any reagents**. Sodium hydroxide, hydrochloric acid, sulfuric acid, iodine, and many other chemicals suggested in this book must be handled with care. Some are also flammable, such as alcohol/ethanol. Alcohol, hexane and other solvents, as well as a variety of gases produced in some of the experiments, are flammable.
- **Bunsen flames and flammable liquids** – use the yellow safety flame, or turn the Bunsen burner off when not in use. Make sure the Bunsen flame is out before handling flammable liquids, such as alcohol/ethanol, hexane, cyclohexane and some other solvents. An alternative may be to heat water using a kettle or use a water bath.

Explosive
These substances, if treated incorrectly, may explode. Explosive substances must be handled very carefully.

Flammable
These substances can easily catch fire.

Oxidising
These substances provide oxygen, which allows other materials to burn more fiercely.

Gas under pressure
The container contains pressurised gas. This may be cold when released or explosive when heated. Container should not be heated.

Acute toxicity
These substances can cause death. Avoid skin contact and ingestion.

Corrosive
These substances attack or destroy living tissues, including eyes and skin.

Moderate hazard
These substances may irritate your skin or eyes, or may show mild toxicity.

Health hazard
Can cause serious long-term health effects from short- or long-term exposure. Avoid skin contact and ingestion.

Hazardous to the aquatic environment
These substances are toxic to aquatic life and may cause long-term environmental effects. They should be disposed of correctly.

Figure 1 International hazard warning symbols

International hazard warning symbols

You will need to be familiar with these symbols when undertaking practical experiments in the laboratory. Make sure you fully understand the hazards indicated (Figure 1).

The bottles of chemicals you will be using will have one or more of these symbols on them, like the one shown in Figure 2.

Remember, laboratories are safe places if you work carefully, tidily and safely!

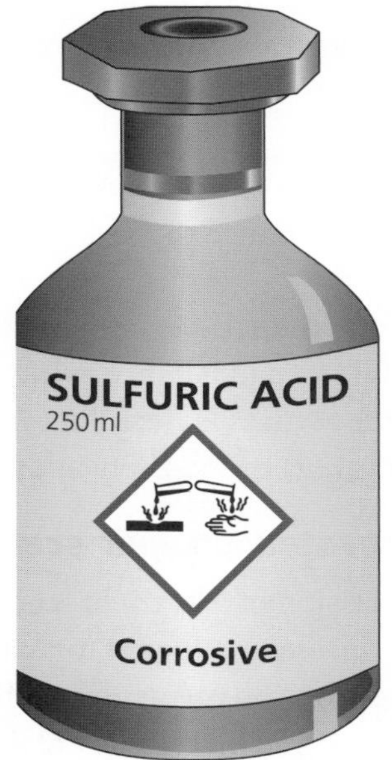

Figure 2 Sulfuric acid is corrosive.

Special note to teachers

In the suggested practical exercises, materials are used which, although familiar in many cases, are of a potentially hazardous nature, and appropriate care and precautions should be taken. We believe that the experiments can be carried out safely in school laboratories. However, it is the responsibility of the teacher to make the final decision depending on the circumstances at the time. Eye protection should be worn at all times. In certain cases, disposable gloves and fume cupboards will be required. Teachers must ensure that they follow the safety guidelines set down by their employers, and a risk assessment must be completed for *any* experiment that is carried out. Teachers should draw students' attention to the hazards involved in the particular exercise to be performed. The hazards are shown within the 'Safety' section of the individual practicals.

It is recognised that, in some cases, there may not be sufficient apparatus to carry out a class practical. If there is insufficient apparatus, perhaps the experiment can be carried out as a demonstration (consider using some student assistance).

In some cases, certain pieces of apparatus may not be available. Use alternatives if possible, **as long as the safety precautions are not overlooked**. For example, burettes, pipettes and gas syringes could be replaced by measuring cylinders of suitable sizes. If you substitute equipment, attention should be drawn to the precision of using the alternative – usually much lower than using that suggested.

It is recommended that certain experiments are carried out in a fume cupboard, see individual practical Safety Guidance notes. If a fume cupboard is not available, an alternative may be to use a well-ventilated room for the experiment or take the experiment outside. However, where relevant, it must be pointed out that the fumes produced are noxious and may cause an asthmatic attack. All the same safety precautions should still be followed.

Plasticine and cocktail sticks can be used as an alternative to molecular models. Marshmallows can also be used but ensure they are thrown away afterwards, and **not** eaten!

Using and organising techniques, apparatus and materials

In an experiment, you will first have to decide on the measurements to be made and then collect together the apparatus and materials required. The quantities you will need to measure most often in laboratory work are **mass**, **length** and **time**.

- What apparatus should you use to measure each of these?
- Which measuring device is most suitable?
- How do you use the device correctly?

Figure 3 A digital top pan balance

Precision of measuring instruments

Make a list of the apparatus you use in an experiment and record the smallest division of the scale of each measuring device. This is the **precision** of your measurements. For example, if the divisions on the scale on a thermometer are at 1 °C intervals, the precision of a temperature reading will be 1 °C.

Balances

A **balance** is used to measure the mass of an object. There are several types available.

- In a beam balance the unknown mass is placed in one pan and balanced against known masses in the other pan.
- In a lever balance a system of levers acts against the mass when it is placed in the pan.
- A digital top pan balance, which gives a direct reading of the mass placed on the pan, is shown in Figure 3.
- The unit of mass is the kilogram (kg).
- The gram (g) is one-thousandth of a kilogram: $1\,g = \frac{1}{1000}\,kg = 10^{-3}\,kg = 0.001\,kg$

How accurately do your scales measure?

- The precision of a beam balance is the size of the smallest mass that tilts the balanced beam.
- The precision of a digital top pan balance is the size of the smallest mass that can be measured on the scale setting you are using, probably 0.01 g.
- When using an electronic balance you should wait until the reading is steady before taking it.

Ruler

- The unit of length is the metre (m).
- Multiples are:
  - 1 decimetre (dm) = 10^{-1} m
  - 1 centimetre (cm) = 10^{-2} m
  - 1 millimetre (mm) = 10^{-3} m
  - 1 micrometre (μm) = 10^{-6} m
  - 1 kilometre (km) = 1000 m

- A **ruler** is often used to measure lengths in the centimetre range.
- The correct way to measure with a ruler is shown in Figure 4, with the ruler placed as close to the object as possible.
- The smallest scale division on a ruler is 1 mm. It may also be possible to read the ruler to the nearest 0.5 mm if the reading is between two of the scale marks.
- The accuracy of the measurement will be greater, the longer the length measured:
  - For a measured length of 1 m = 1000 mm, a measurement error of 1 mm is 1 part in 1000.
  - For a measured length of 1 cm = 10 mm, a measurement error of 1 mm is 1 part in 10.

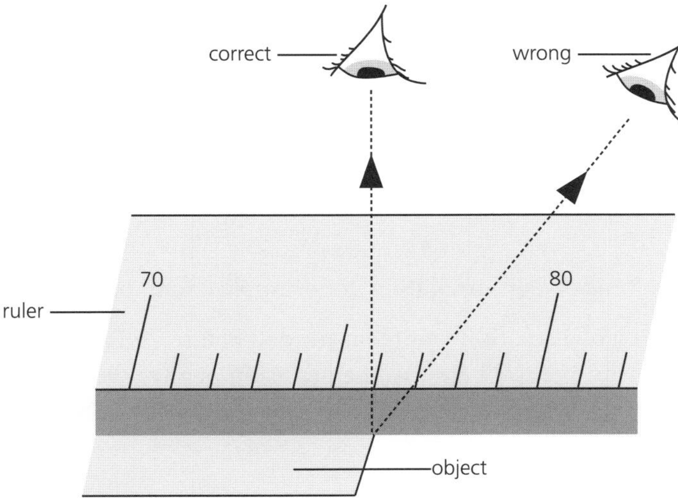

Figure 4 Using a ruler: The reading is 76 mm or 7.6 cm. Your eye must be directly above the mark on the scale or the thickness of the ruler causes parallax errors.

Clocks and timers

- Clocks, watches and timers can be used to measure time intervals. In an experiment it is important to choose the correct timing device for the required measurement.
- The unit of time is seconds (s).
- A stopwatch will be sufficient if a time in minutes or seconds is to be measured, but if times of less than a second are to be determined then a digital timer is necessary (Figure 5).

Figure 5 This stopwatch can be used to measure the time passed in a chemical reaction.

How accurate are your timings?

- When using a stopwatch, human reaction times will influence the reading. Human reaction time varies but is likely to be about 0.5 s. The reading on the stopwatch may be to the nearest 0.1 s or even 0.01 s, but the timing is only accurate to about 0.5 s.
- For time intervals of the order of seconds, a more accurate result will be obtained by measuring longer time intervals – for example, time reaction rate over 60 seconds rather than over 15 seconds.

Changing measurements

- Take readings more frequently if values are changing rapidly.
- It will often be helpful to work with a partner who watches the timer and calls out when to take a reading.
- Pressing the lap-timer facility on the stopwatch at the moment you take a reading freezes the time display for a few seconds and will enable you to record a more accurate time measurement.

Measuring cylinders

- The volume of a liquid can be obtained by pouring it into a **measuring cylinder** (Figure 6).
- Measuring cylinders are often marked in millilitres (ml) where 1 millilitre = 1 cm^3.
- The accuracy of the reading depends on the smallest scale division. For example, this could be 10 cm^3 or 1 cm^3, depending on the size of the measuring cylinder.
- Note that 1 litre = 1000 cm^3 = 1 dm^3.

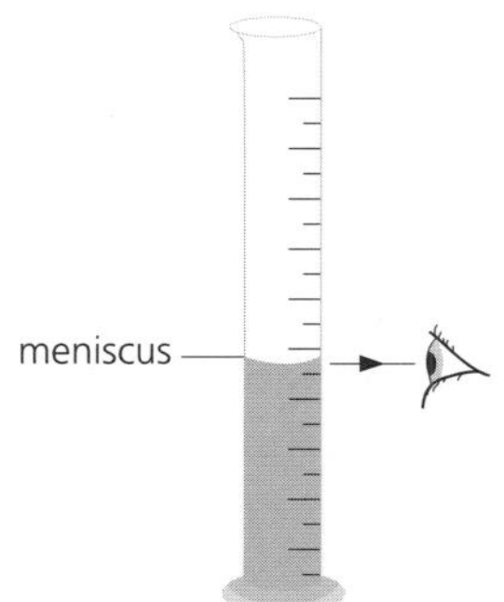

Figure 6 When making a reading the measuring cylinder should be vertical and your eye should be level with the bottom of the curved liquid surface – the meniscus.

Measuring volume accurately

A beaker or measuring cylinder is often used when the volume measurement is not needed to be very accurate. When carrying out titrations, pipettes and burettes are used to give a more precise volume reading.

Pipettes are often used to accurately measure a fixed volume of a given solution, for example 25 cm^3 or 50 cm^3.

Burettes are used in titrations where an accurate reading of a solution volume is needed before and at the end of the titration. The smallest scale division on a burette is 0.1 cm^3, so the precision of any volume measured with the burette will be 0.1 cm^3. It may also be possible to read the burette to the nearest 0.05 cm^3 if the bottom of the meniscus is between divisions.

Observing, measuring and recording

Having collected together and familiarised yourself with the equipment and materials needed for an experiment, you are now ready to start making some observations and measurements.

- You should also record any difficulties encountered in carrying out the experiment and any precautions taken to achieve accuracy in your measurements.
- Do not dismantle the equipment until you have completed the analysis of your results and are sure you will not have to repeat any measurements!
- How many significant figures will your data have? What are the colour changes and other observations?
- How will you record your results?

Significant figures

- The number of digits given for a value in a measurement or calculated value indicates how accurate we think it is. This is called the number of **significant figures**. A measurement written as 5 cm^3 has one significant figure but a measurement written as 5.0 cm^3 has two significant figures. The 0 after the decimal point is significant because it shows we have measured a volume to the nearest 0.1 cm^3.
- When doing calculations, your answer should have the same number of significant figures as the measurements used in the calculation. For example, if your calculator gives an answer of 1.23578, this would be 1.2 if your measurements have two significant figures and 1.24 if your measurements have three significant figures.
- To round a number to two significant figures, look at the third digit. Round up if the digit is 5 or more, and round down if the digit is 4 or less. So 1.23 rounds to 1.2 but 1.25 rounds to 1.3.
- If a number is expressed in standard notation, the number of significant figures is the number of digits before the power of 10. For example, 6.24×10^2 has three significant figures.
- If values with different numbers of significant figures are used to calculate a quantity, quote your answer to the smallest number of significant figures.

Sources of error

Every measurement of a quantity is an attempt to find its true value and is subject to errors arising from the limitations of the apparatus and the experimental procedure.

Systematic errors

An error that is introduced by the system is called a systematic error. Sources of systematic error can include the:

- environment you are working in
- methods of observation
- instruments used – each will have its own source of error.

For example:

- If the temperature at which you make a measurement is 10 °C higher than the temperature in another laboratory where the same experiment is performed, this could cause your measurements to be consistently different.
- When reading a burette volume you should ensure that your eye is directly opposite the bottom of the meniscus, or your measurements of volume will consistently be too low or too high as shown in Figure 7. Also is the burette vertical? Is the volume of liquid in the burette at the zero mark before you start the titration? If not then the reading will be in error. Another source of error when using a pipette is to fail to allow enough time for a pipette to drain properly. This produces an error in the volume you are measuring out!
- Instrument errors include when a top pan balance or thermometer is not calibrated correctly.

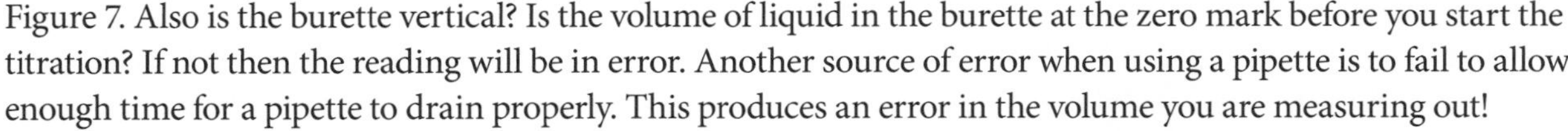

Figure 7 A systematic error created by not viewing the bottom of the meniscus on a burette.

Random errors

Random errors usually result from not taking the same measurement in exactly the same way – for example judging when a colour change has occurred. This type of error produces readings that vary randomly above or below the true value.

Some ways of reducing random errors are:

- Taking repeated measurements so that you can obtain an average value.
- Plotting a graph to establish a pattern and then drawing a line or curve of best fit.
- Maintaining good experimental technique, for example measuring volume of solutions carefully.

Tables

If several measurements of a quantity are being made, draw up a **table** in which to record your results.

- Use the column headings, or start of rows, to **name** the measurement and state its unit.
- For example, the reaction of magnesium and hydrochloric acid was observed when varying the temperature by recording the time it took for the piece of magnesium ribbon to disappear. The results were represented in a table (Table 1).
- Repeat the measurement of each observation if possible and record the values in your table. If repeat measurements for the same quantity are significantly different, take a third reading. Calculate an average value from your readings. When you calculate an average from your results it is very important that you do not use any anomalous results to calculate the average.
- Numerical values should be given to the number of significant figures appropriate to the measuring device.

Table 1 A data table for the study of the variation of rate of reaction between magnesium and hydrochloric acid with temperature.

Temperature / °C	Time taken for the magnesium ribbon to disappear / seconds			
	1	2	3	Average
20	50	52	51	51
25	28	29	30	29
30	27	26	25	26
35	21	23	22	22
40	15	13	14	14
45	10	8	9	9

Handling experimental observations and data

Now that you have collected your measurements, you will need to process them. Perhaps there are calculations to be made or you may decide to draw a graph of your results.

Then you can summarise what you have learnt from the experiment, discuss sources of measurement error and draw some conclusions from the investigation.

- What is the best way to process your results?
- Are there some inconsistent measurements to be dealt with? **Anomalous data** are readings which fall outside the normal, or expected, range of measurements. In the study of the loss of mass of calcium carbonate lumps when heated with time the anomalous results are clearly seen (Figure 8).
- What sources of errors are there?
- What conclusions, generalisations or patterns can you draw?

Calculations

You may have to produce an average (mean) value to process your results.

The **mean** is found by taking all the measurements you have made, adding them together and dividing by the number of measurements taken. When calculating an average, only use concordant values; these are repeated results which are similar within certain limits of experimental accuracy.

For example, in a titration if the four burette readings obtained were 23.8 cm^3, 23.9 cm^3, 24.4 cm^3 and 23.9 cm^3 the average would be found by initially realising the 24.4 cm^3 result is not a concordant result, and that it should not be used in calculating the average value.

$$\text{The average} = \frac{23.8 + 23.9 + 23.9}{3} = 23.9\,\text{cm}$$

The value has been given to three significant figures because the precision of the burette was to 0.1 cm^3.

Graphs

Graphs can be useful in finding the relationship between two quantities. Graphs are a pictorial way of looking at data from a table. Certainly, trends can be observed and data that is out of sequence with the rest can be seen. When drawing graphs, ignore anomalous data.

- You will need at least six data points taken over as large a range as possible to plot a graph, in order to see a trend.
- Choose scales that make it easy to plot the points and use as much of the graph paper as possible.
- Make sure you label each axis of the graph with the name and unit of the quantity being plotted.
- Mark the data points clearly with a dot in a circle ⊙ or a cross (×) using a sharp pencil.
- Join up your points with a smooth line or curve of best fit.

Any anomalous points are clearly seen when smooth curves are drawn.

- Always draw straight lines (Figure 9) and curves of best fit when producing graphs.
- Note that the line must go through the origin for the quantities to be proportional.
- If a straight-line graph does not go through the origin, one can only say that there is a linear dependence between y and x.

Errors

- In practice, points plotted on a graph from actual measurements may not lie exactly on a straight line or curve of a graph due to measurement errors.
- The 'best straight line' is then drawn through the points, as mentioned earlier.
- If possible, repeat any anomalous measurements to check that they have been recorded properly or try to identify the reason for the anomaly.

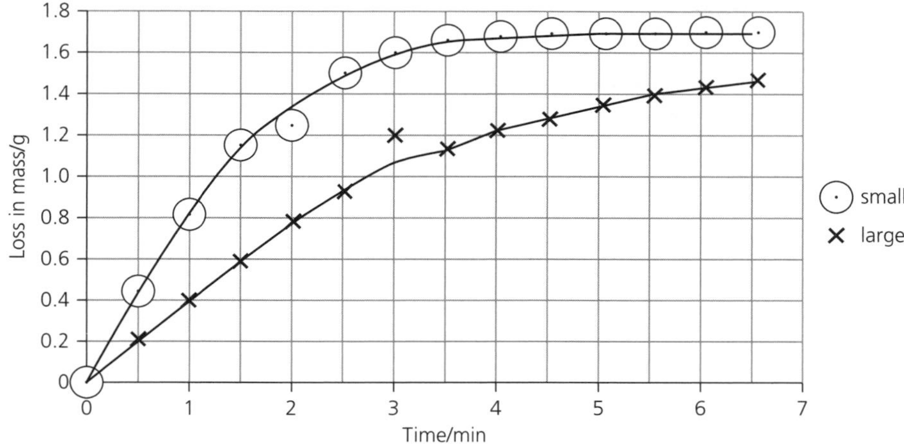

Figure 8

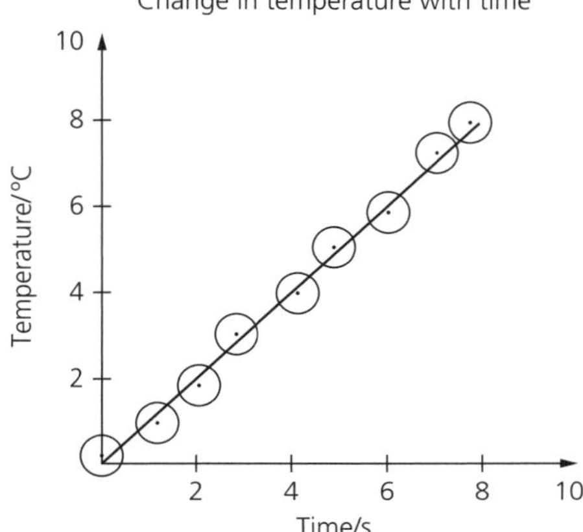

Figure 9 A line of best fit has been drawn to show an accurate trend for the data plotted.

Gradient

Gradients are very important in interpreting the results obtained from an experiment, especially those involving rates of reactions.

Figure 10 shows a typical graph obtained when magnesium metal reacts with hydrochloric acid to give hydrogen gas.

To help interpret this graph in terms of the rate of reaction you need to look at the steepness of the curve as the reaction progresses. The steeper the curve the faster the rate. You should be able to see that the steepness, or gradient, of the curve decreases with time and eventually the reaction stops after 110 s when the graph becomes flat. What your mind is actually doing to allow it to come to this conclusion is looking at the slope of the tangent to the curve at different points along the curve.

A tangent is a line that touches the curve at a specific point, but would not cross the curve if it were extended either way.

To draw a tangent, place a ruler at the point on the curve where you wish to find the rate. Adjust the angle of the ruler so that it follows the slope of the line at that point.

On Figure 10 it is easy to see that the tangent drawn at 20 s is much steeper than that drawn at 60 s. This allows us to say that the rate of reaction is faster at 20 s than at 60 s.

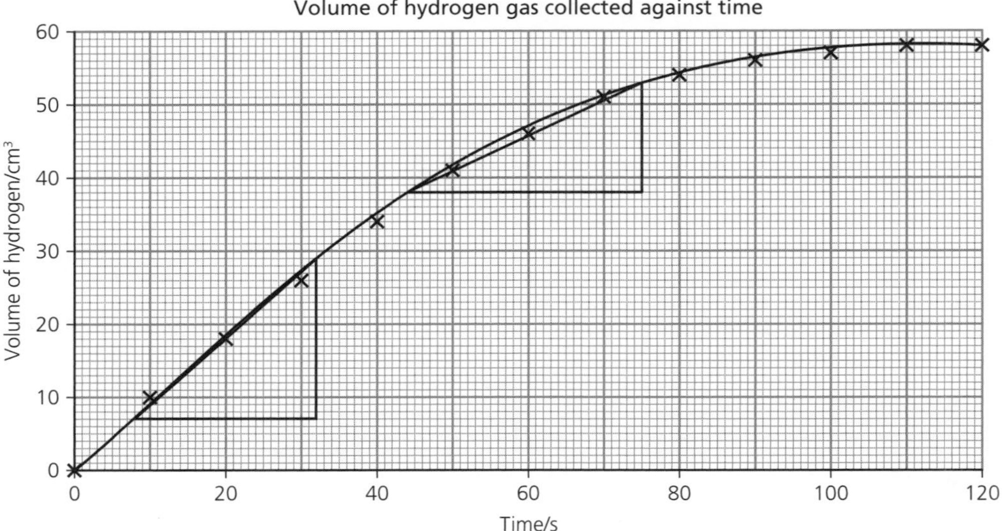

Figure 10

To calculate the rate of the reaction after 40 seconds you would first draw a tangent at 40 seconds and then find the gradient by constructing the triangle as shown in Figure 11. It is better to draw a large triangle as this will give a more accurate measurement of the rate.

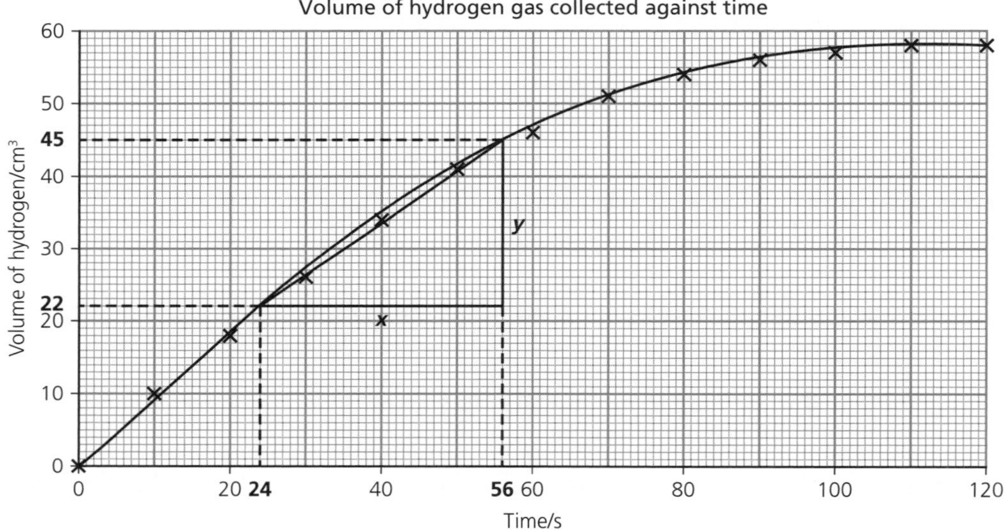

Figure 11

The units of the gradient are found from the units on the axes. In Figure 11, using the dashed lines:

$$\text{The slope of the tangent, the rate} = \frac{(\text{value of } y)}{(\text{value of } x)}$$

$$= \frac{(45\,\text{cm}^3 - 22\,\text{cm}^3)}{(56\,\text{s} - 24\,\text{s})} = 0.72\,\text{cm}^3/\text{s}$$

This method gives a numeric value for the rate at 40 seconds after the start of the experiment of $0.72\,\text{cm}^3/\text{s}$.

Conclusions

Once you have analysed your experimental results, summarise your conclusions clearly and relate them to the aim of the experiment.

- State whether a hypothesis has been verified. If your results do not, or only partially, support a hypothesis, suggest reasons why.
- If a numerical value has been obtained, state it to the correct number of significant figures. Compare your results with known values, if available, and suggest reasons for any differences.
- State any relationships discovered or confirmed between the variables you have investigated.
- Mention any patterns or trends in the data.

Evaluating investigations

Finally, evaluate the experiment and discuss how it could be improved. Could some things have been done better? If so, suggest changes or modifications that could be made to the procedure or the equipment used in the investigation. For example:

- Should repeat measurements be made?
- Are there enough results to show a pattern?
- Was the range of the independent variable good enough?
- Should you get further data in between values – for example if there is an uncertainty about the results in one part of the range?
- Identify and comment on sources of error in the experiment. For example, it may be very difficult to eliminate all energy losses to the environment in an experiment where the temperature change of a liquid is measured; if that is the case, say so.
- Mention any sources of systematic error in the experiment, or random errors and what might have caused them.

Planning investigations

- When preparing a plan to answer a specific question or extend a method to a new situation, you should produce a logical and safe procedure.
- Identify the variables in the investigation and decide which ones to investigate and which ones you should try to keep constant so that they do not affect the results. The variable that is changed is known as the **independent variable**. The variable that is measured is known as the **dependent variable**. To discover the relationship between variables, you should change only one variable at a time.
- Once you know what you need to measure, you can decide on the apparatus and materials to be used. You should ensure that you choose measuring devices that have sufficient precision.
- Before you write a plan, familiarise yourself with how to use the apparatus and develop a plan of work. It will be helpful to decide how to record your results; draw up tables in which to record your measurements if appropriate.
- Describe how you would carry out the experiment, and include any safety precautions. It is useful to include a sketch of the experimental set-up (Figure 12, for example).

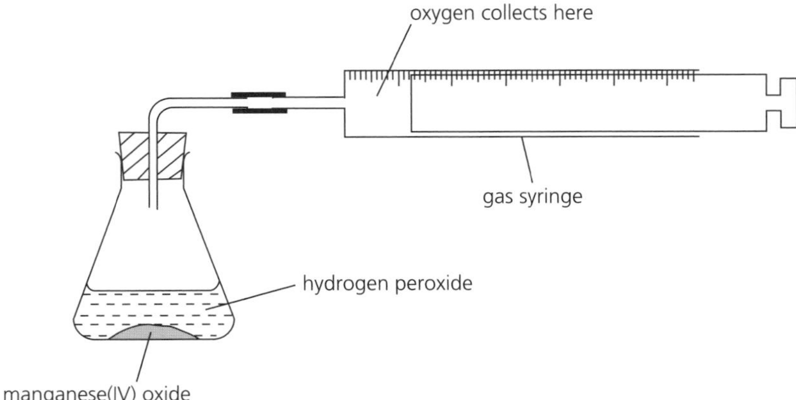

Figure 12 This apparatus can be used to investigate how the volume of oxygen produced varies with the concentration of hydrogen peroxide, using manganese(IV) oxide as a catalyst.

1 States of matter

1.1 Rate of diffusion of ammonia and hydrogen chloride (Teacher demonstration)

When two or more gases come into contact with each other they slowly mix together. Diffusion is the movement of gases from an area of high concentration to a lower concentration. In this experiment hydrogen chloride gas, HCl(g), evaporates from concentrated hydrochloric acid. Ammonia gas, $NH_3(g)$, evaporates from a concentrated ammonia solution. The two gases will mix and react to form the white solid ammonium chloride.

ammonia + hydrogen chloride → ammonium chloride

$NH_3(g) + HCl(g) \rightarrow NH_4Cl(s)$

This is a teacher demonstration and it is recommended that it is performed in a fume cupboard. It is also recommended that a sheet of black paper be placed behind the demonstration so that the formation of the white solid ammonium chloride can be observed clearly.

KEY TERMS 

diffusion
kinetic particle theory
relative molecular mass

Aim

To determine how the mass of gas particles affects the rate of diffusion of gases.

Apparatus and chemicals

- Eye protection
- Access to a fume cupboard
- Protective gloves
- Stopwatch
- A length of glass tubing of recommended length 50–100 cm with internal diameter of 2–3 cm
- 2 × retort stands complete with bosses and clamps
- 2 × cotton wool buds
- 2 × bungs to fit the ends of the glass tube
- Concentrated hydrochloric acid
- Concentrated ammonia solution

SAFETY GUIDANCE 

- Eye protection must be worn.
- This demonstration must be setup and carried out in a fume cupboard.
- Care should be taken when opening the bottle of the ammonia solution because pressure can build up inside the bottle. Do not keep the bottle of ammonia open for a long period of time.
- Concentrated ammonia solution – corrosive, hazardous to the aquatic environment.
- Ammonia gas – toxic, hazardous to the aquatic environment.
- Concentrated hydrochloric acid – corrosive.
- Hydrogen chloride gas – corrosive.
- Ammonium chloride – low hazard.

Method

Throughout, the teacher/demonstrator must wear eye protection and safety gloves.

1 While working in a fume cupboard, clamp the glass tubing at both ends. Ensure that the glass tubing lies horizontally.

2 Open the bottle of concentrated ammonia solution and leave it in the fume cupboard. Point the bottle away from yourself and the students. Then open the bottle of concentrated hydrochloric acid, again in the fume cupboard, and hold the stopper near the mouth of the bottle of ammonia solution. Note the white clouds of ammonium chloride that are produced.

3 Put the end of one of the cotton buds into the concentrated ammonia solution. Quickly repeat this procedure with the second cotton bud and the concentrated hydrochloric acid. Push the buds into opposite ends of the tube and start a stopwatch. Replace the tops on the bottles of the ammonia and hydrochloric acid solutions.

TIP

Your teacher will perform the demonstration but you should read through the method to ensure you understand it and so you can answer the questions about the method.

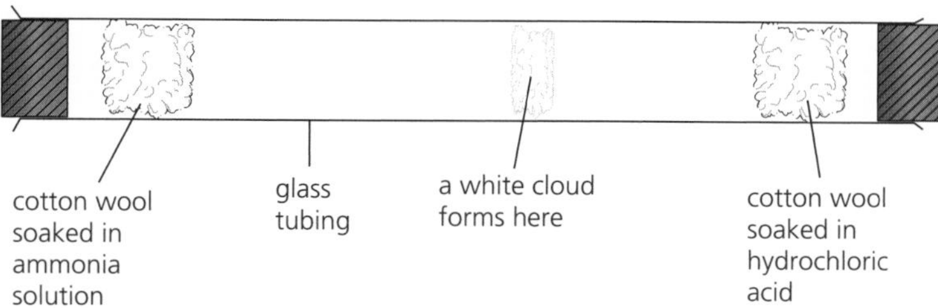

Figure 1 Diagram of the apparatus

4 Watch the ring of white solid forming in the tube, and stop the stopwatch when it is initially observed. Record the time in the Observations section.

5 Measure the distance each gas has travelled from its end of the tube to where the white solid formed. Record these distances in the Observations section.

Observations

Time taken for the white solid to form: .... seconds

Distance travelled by ammonia gas: .... cm

Distance travelled by hydrogen chloride gas: .... cm

Conclusions

1 How did the two gases travel along the tube?

....

2 Use your results to find the rate at which the two gases diffused along the tube:

Rate of diffusion of the ammonia gas: .... cm/s

Rate of diffusion of the hydrogen chloride gas: .... cm/s

3 Why is it important that the glass tube is clamped horizontally before the experiment begins?

TIP

Rate is how quickly something happens. Here we are not calculating the rate of a chemical reaction but the rate of diffusion of the two gases.

....

4 Why are the solutions which produce the gases placed at the end of a glass tube, rather than just placing them at opposite sides of a closed fume cupboard?

....

5 Why is it important that the experiment takes place in a fume cupboard?

....

6 Suggest how increasing the length of the tube would affect the time of formation of the white solid.

....

....

....

....

7 Suggest what other variables would affect the time of formation of the white solid.

....

....

....

....

8 Does the ring form closer to the ammonia or hydrogen chloride end of the tube?

....

9 Explain why the white solid forms closer to this end of the tube.

....

....

Evaluation

Discuss how the procedure could be improved to give stronger evidence for your answers in questions 1 and 8.

....

....

....

GOING FURTHER

Do you think that the rate of diffusion of the ammonia and hydrogen chloride gases you have calculated is the same as the speed of the particles (molecules) of the gases? Explain your answer.

....

....

....

....

....

1.2 Reaction of potassium iodide with silver nitrate

When potassium iodide and silver nitrate react together, one of the products formed is silver iodide, which is a yellow-coloured compound. This colour change can be used to indicate that a chemical reaction has occurred.

KEY TERM

kinetic particle theory

potassium iodide + silver nitrate → potassium nitrate + silver iodide

$$2KI + AgNO_3 \rightarrow 2KNO_3 + AgI$$

In the experiment you will perform this reaction with both solid reactants and then again with solutions.

Aim

To discover how the speed at whicн a reaction occurs between solids compares to the same reaction using solutions.

Apparatus and chemicals

- Eye protection
- Test tubes
- 2 × small spatulas
- 2 × dropping pipettes
- Rubber bung for boiling tube
- Test-tube rack
- 2 × 1 cm^3 dropping pipettes
- Small amount of silver nitrate solid and 0.05 mol/dm^3 solution
- Potassium iodide solid and 0.1 mol/dm^3 solution

SAFETY GUIDANCE 

- Eye protection must be worn.
- Silver nitrate – corrosive, oxidising, toxic.
- Silver iodide – toxic, hazardous to the aquatic environment.
- Potassium nitrate – oxidising.
- Potassium iodide – low hazard.

Method 1 (Teacher demonstration)

Throughout the practical the teacher and student should wear eye protection.

1. The teacher mixes a quarter of a spatula of potassium iodide with a quarter of a spatula of silver nitrate in a test tube.
2. The teacher puts a rubber bung into the test tube.
3. Taking care not to be near any solid surface, the teacher shakes the test tube for a few minutes.
4. After shaking it, the teacher can show the test tube to the class.
5. Write down your observations in the Observations section.

Method 2

Throughout the practical the student should wear eye protection.

1. Using one of the dropping pipettes, place 1 cm^3 of potassium iodide solution into the test tube.
2. Using the other dropping pipette, add 1 cm^3 of silver nitrate to the same test tube.
3. Place the test tube into the rack.
4. After 5 minutes write down your observations in the Observations section.

Observations

Observation when solid silver nitrate is shaken with solid potassium iodide.

....

....

Observation when solutions of silver nitrate and potassium iodide are mixed together.

....

....

....

....

What colour is potassium nitrate solution?

....

Conclusions

1. In Method 1, explain why the test tube was shaken after the two solids had been added to it.

....

....

....

2 In Method 1, why was it important that the teacher was not near any solid surfaces before starting to shake the test tube?

....

....

3 In both procedures why were different spatulas and different pipettes used to measure out each substance?

....

....

4 What was observed in both Methods 1 and 2 which told you that a chemical reaction had occurred?

....

5 Which of the two reactions occurred faster?

....

6 Use ideas about particles in solids and solutions to explain why the reaction you have stated in point 5 occurred faster.

....

....

....

Evaluation

1 In any investigation there should only be one variable altered – the one you are investigating. What could have been done to ensure that this was true?

....

....

2 Suggest how the reaction between solid reactants used in Method 1 could be speeded up. Explain why the change you have suggested would speed up the reaction.

....

....

....

1.3 Determination of the freezing point of stearic acid

Stearic acid is a solid at room temperature. When heated it melts to form a liquid. In this experiment you will heat stearic acid in a water bath past its melting point. You will then allow it to cool for 15 minutes, taking the temperature at regular intervals as it forms a solid again.

KEY TERM 

melting point

You will use the data you obtain to plot a graph of temperature against time (a cooling curve) and find the freezing point of stearic acid.

Aim

To determine the freezing point of stearic acid.

Apparatus and chemicals

- Eye protection
- 250 cm^3 beaker
- Boiling tube half full of solid stearic acid
- Thermometer
- Boiling-tube rack
- Kettle (for hot water)
- Stopwatch
- Tongs

SAFETY GUIDANCE

- Eye protection must be worn.
- Take care when handling the hot glassware and hot liquids. Use tongs and avoid direct contact with the boiling tube until it is cooled, to avoid burns. Do not attempt to move the beaker of hot water.
- You should stand while carrying out the experiment and making measurements so that you can move away instantly if the beaker or boiling tube containing the hot liquid breaks or is knocked over.
- Avoid getting the stearic acid on your skin.

Method

Throughout the practical the student should wear eye protection.

1 Place the boiling tube with stearic acid into a beaker. Into the beaker, pour water that has been heated to around 80 °C, which is above the melting point of stearic acid.

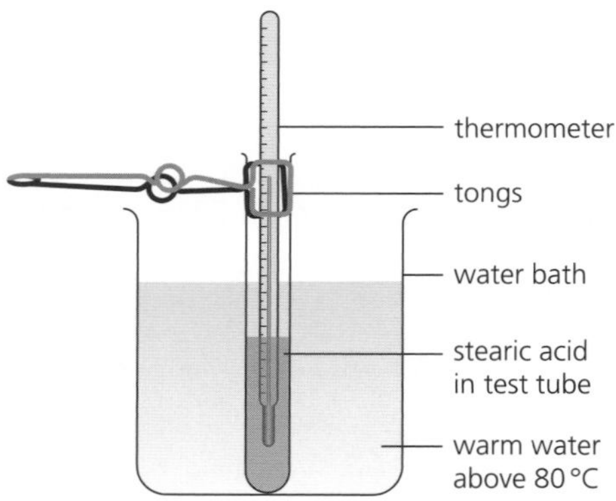

Figure 2

2 Once the stearic acid has melted, remove the boiling tube from the beaker of water using the tongs, and place the tube in the boiling-tube rack.

3 Place the thermometer in the tube. Measure the temperature of the stearic acid every minute for 15 minutes using the thermometer to stir the stearic acid whilst it is still a liquid. Record the time and temperature in the table in the Observations section.

4 At the end of the experiment you will need to heat the stearic acid to melt it once more to remove the thermometer. To do this, place the boiling tube back into a beaker of boiling water from a recently boiled kettle. Remove the thermometer when the stearic acid has once again melted.

Observations

Time / min	Temperature /°C
0	
1	
2	
3	
4	
5	
6	
7	

Time / min	Temperature /°C
8	
9	
10	
11	
12	
13	
14	
15	

1 Draw and label axes for plotting this data with the temperature of stearic acid on the *y*-axis and time on the *x*-axis.

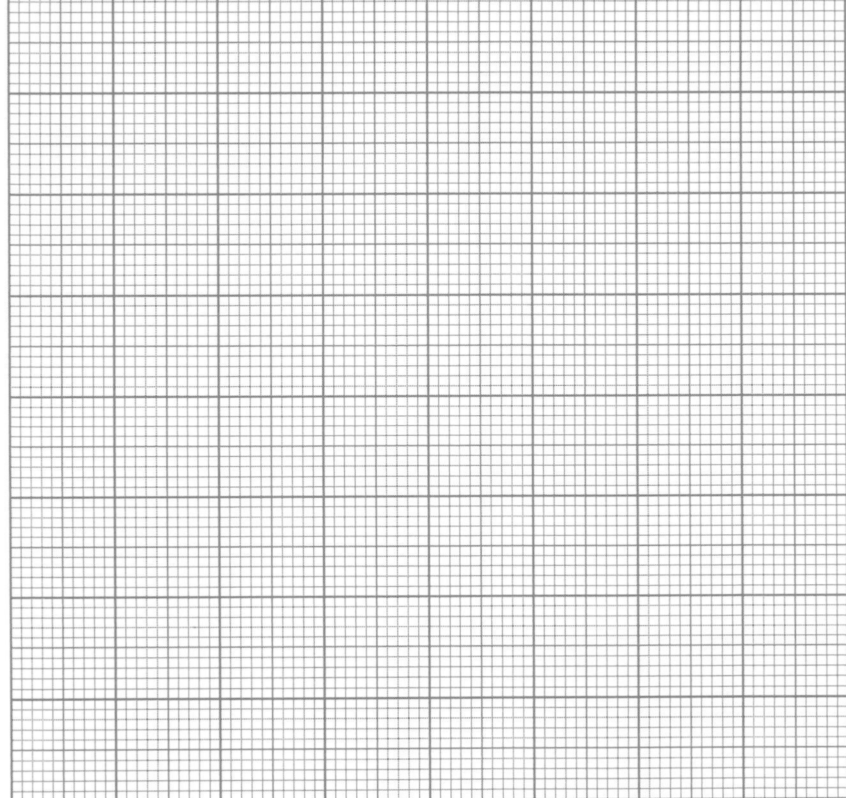

TIP

Remember to choose a suitable scale for the axes and make sure you label the axes, including the units.

2 Plot the points and draw a line of best fit.

Conclusions

1 At what temperature did the stearic acid begin to change state?

....

2 How could you tell this from your graph?

....

3 What is happening to the molecules of stearic acid in the first part of the graph?

....

....

4 Explain the shape of your graph in terms of what is happening to the molecules of stearic acid at the different stages.

....

....

....

....

....

5 Describe what you would expect to observe if you continued monitoring the temperature for another 30 minutes. Give a reason for your answer.

....

....

....

Evaluation

1 State why it was important to remove the boiling tube with the stearic acid from the water.

....

....

2 Explain why it was important to stir the stearic acid with the thermometer.

....

3 Explain why it is important to take temperature readings every minute, rather than measuring the temperature when you first observe that the liquid stearic acid changes to a white solid.

....

4 Describe how this experiment could be improved to make a more accurate measurement of the freezing point of stearic acid.

....

....

....

....

GOING FURTHER

A student repeated the experiment but placed the tube of molten stearic acid in a polystyrene cup containing a small amount of cold water. The student measured both the temperature of the stearic acid and the temperature of the water every minute for 15 minutes.

The water temperature increased as the stearic acid cooled, showing that energy was being transferred from the hot tube to the water in the cup. The student noticed, however, that while the temperature of the stearic acid did not change as it changed from liquid to solid, the water temperature continued to rise.

Explain why the water temperature continued to increase.

....

....

....

....

....

2 Atoms, elements and compounds

2.1 Elements, mixtures and compounds

In this experiment the metal element iron and the non-metal element sulfur will be mixed and then reacted together to form the compound iron(II) sulfide.

iron + sulfur → iron(II) sulfide
Fe(s) + S(s) → FeS(s)

KEY TERMS

element
mixture
compound

The elements iron and sulfur, the mixture of these two elements and the iron(II) sulfide produced will then be reacted with water and dilute hydrochloric acid. Also, the effect of a magnet will be demonstrated. Using the results of these experiments, the properties shown by the elements, mixtures and compound will be identified.

Part 1 is a teacher demonstration and it is recommended that it is performed in a fume cupboard, or at least in a well-ventilated laboratory with a safety screen.

Part 2 is the observations and test-tube reactions students make during the demonstration.

Aim

To identify the differences between elements, their mixtures and compounds.

SAFETY GUIDANCE

- Throughout, the teacher/demonstrator/students must wear eye protection.
- Toxic fumes are produced in the experiment (point 6 of method) and so a fume cupboard is recommended, but if that is not available then a well-ventilated laboratory is needed and the use of a safety screen would then be essential for the demonstration.
- Iron filings – highly flammable
- Sulfur – low hazard
- Hydrochloric acid (2 mol/dm^3) – irritant
- Iron(II) sulfide – low hazard
- Hydrogen sulfide – a toxic foul-smelling gas is produced in the demonstration. Use a fume cupboard when adding acid to the iron(II) sulfide.

TIP

You can test the conductivity of graphite by sharpening both ends of a graphite pencil.

Apparatus and chemicals

- Eye protection
- Protective gloves
- Spatula
- 9 × heat-resistant test tubes
- Heat-resistant mat
- Bunsen burner
- Test-tube holder
- Test-tube rack
- Small bar magnet
- Access to fume cupboard
- Powdered sulfur
- Fine iron filings
- Water
- 2 mol/dm^3 hydrochloric acid
- Paper towels
- Mineral wool

Method 1 – teacher demonstration

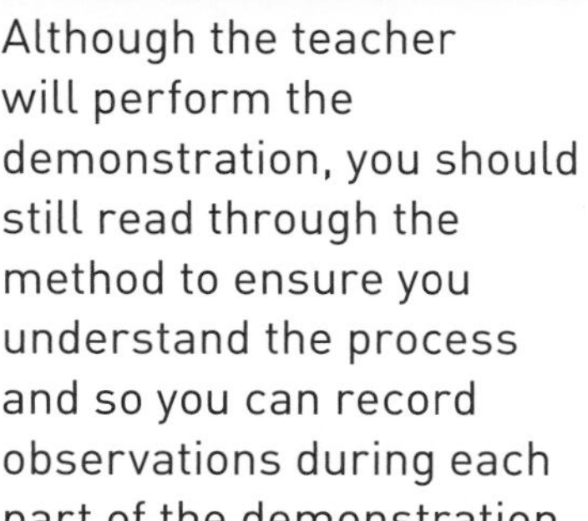

TIP
Although the teacher will perform the demonstration, you should still read through the method to ensure you understand the process and so you can record observations during each part of the demonstration.

Eye protection must be worn by the teacher/demonstrator/students.

1 Initially, separately test the iron filings and sulfur powder with a magnet wrapped around by a paper towel, by putting it very close to the iron filings and the sulfur powder.

2 One spatula measure of iron filings with one spatula measure of sulfur powder on a paper towel are then mixed and tested with a magnet, as in step 1.

3 Mix 1 g of iron filings with 1 g of sulfur and heat strongly in a heat-resistant test tube that has a mineral wool plug in it. This should be done until the red glow disappears.

4 Put the test tube aside, on a heat-resistant mat, to cool. Turn the Bunsen burner off.

5 When the test tube has cooled, tip the contents onto a paper towel and re-test with a magnet.

6 In a fume cupboard, test iron(II) sulfide with water and dilute hydrochloric acid.

Method 2 – observations and student practical

1 Put on your eye protection.

2 Record in the table:
   a the appearance of the iron (element) filings, sulfur (element) powder and the mixture produced by your teacher.
   b the effect of a magnet on these substances.

3 Your teacher will carry out a reaction between the iron filings and the sulfur powder.

4 Record in the table:
   a the appearance of the product (iron(II) sulfide) produced by your teacher.
   b the effect of a magnet on this substance.

5 Your teacher will give you six test tubes and a test-tube rack.

6 Into two test tubes add a small amount of iron filings. In another two test tubes add a small amount of sulfur powder and in a third pair add a small amount of the iron/sulfur mixture.

7 To a sample of the iron, sulfur and the iron/sulfur mixture add a little water. Record in the table what happens.

8 In a fume cupboard repeat step 7 with the remaining three samples, but this time use dilute hydrochloric acid. Note down in the Observations table what happens.

9 Your teacher will test the iron(II) sulfide in a fume cupboard. Record in the table what happens.

Observations

	Sulfur	**Iron**	**Mixture**	**Compound (iron(II) sulfide)**
Appearance				
Effect of magnet				
Effect of water				
Effect of dilute hydrochloric acid				

Conclusions

Mixtures behave as the individual .... they are made up from. There is a red glow produced when the .... is heated. This tells us that a .... reaction is taking place and a new substance, a ...., is produced. Compounds differ from the elements or .... in that they are new .... and so behave ....

Evaluation

Outline how this experiment could be improved, or made more reliable.

....

....

....

....

GOING FURTHER

Glucose belongs to a family of compounds called sugars. Sugars are very important compounds. Use your research skills to find out:

a the elements present in glucose ....

b the chemical formula for glucose ....

c why it is an important compound.

....

....

....

3 Bonding and structure

3.1 Properties of ionic and covalent substances

Ionic and covalent substances have different properties.

For ionic compounds:

- They are usually solids at room temperature, with high melting points. They also have high boiling points.
- They are usually crystalline and hard substances.
- They usually cannot conduct electricity when solid, but they will conduct electricity when in aqueous solution or in the molten state.

KEY TERMS 

ionic
covalent

For covalent compounds:

- As simple molecular substances, they are usually gases or liquids at room temperature.
- Generally, they do not conduct electricity either in the solid, liquid or gaseous state, or when dissolved in water.

You are going to investigate some of these properties and identify which of the substances placed at different stations around the laboratory are ionic and which are covalent.

Aim

To identify ionic and covalent types of substances by observing their properties.

Apparatus and chemicals

- Eye protection
- Stoppered bottles labelled A, B, C, D, E, F, G, H

SAFETY GUIDANCE 

- Eye protection must be worn.
- A, B, C, E – low hazard
- D – highly flammable, harmful
- F – toxic, flammable
- G – harmful, irritant
- H – low hazard

Method

1 Put on your eye protection.

2 Your teacher will tell you which substance to start with. Write down its appearance in Table 1.

Do not remove the stopper.

3 Use the appearance, and the data for the melting point and boiling point shown in Table 1, to suggest whether the sample is an ionic or covalent substance (conclusion).

4 Repeat this process with the other seven samples you find around the laboratory.

TIP 

Even though you will not open the bottles you should still wear your eye protection in case the glass breaks.

Observations

Table 1

Substance	Appearance	Melting point /°C	Boiling point /°C	Conclusion
A		−218.80	−83.00	
B		186.00	decomposes	
C		801.00	1413.00	
D		−45.00	156.00	
E		771.00	1500.00	
F		−114.00	78.00	
G		614.00	1382.00	
H		−0.5	101.5	

Conclusion

Substances .... are ionic while substances .... are covalent.

Evaluation

1 What other property could be tested to check your answer as to whether the substance was ionic or covalent?

....

....

2 Describe how you would carry out this test, for the solid and the liquid samples.

....

....

....

....

3 Amina thinks that sample H might be water. Give two reasons why Amina might think this.

....

....

4 Zachariah says that sample H cannot be water. State why Zachariah might say this.

....

....

5 Describe a further test or tests that could help identify sample H.

....

....

....

GOING FURTHER

1 Use your research skills to find out about a technique that can be used to determine the crystal structure of ionic structures.

....

2 Briefly explain how this technique works.

....

....

....

....

3.2 Using electrical conductivity to identify ionic and covalent substances

Ionic substances do not conduct electricity when in solid form. Ionic substances do conduct electricity when in aqueous solution.

The forces of attraction between the ions are weakened and the ions are free to move to the appropriate electrode. This allows an electric current to be passed through the aqueous solution.

Generally, covalent substances do not conduct electricity when in solid form (or liquid) or dissolved in water (if they dissolve in water!)

This is because they do not contain ions.

KEY TERMS 

covalent
ionic
electrode
electrical conductivity
ion

Aim

To test electrical conductivity to classify substances and their solutions as ionic or covalent.

Apparatus and chemicals

- Eye protection
- 2 × carbon electrodes
- Electrode clip holder
- 100 cm^3 beaker
- Crucible
- Spatula
- 2.5 V lamp in a lamp holder
- Three leads plus crocodile clips
- 6 V d.c. power supply or 6 V battery
- Distilled/deionised water
- Strips of polythene
- Candle wax
- Sugar
- Copper(II) sulfate solid
- Magnesium sulfate solid
- 0.2 mol/dm^3 solutions of:
  - Sugar
  - Magnesium sulfate
  - Copper(II) sulfate

SAFETY GUIDANCE 

- Eye protection must be worn.
- Take care when using electrical equipment in a laboratory setting where water is being used.
- Magnesium sulfate – low hazard.
- Copper(II) sulfate: as a solid is toxic, dangerous to the environment; whilst in solution it's corrosive.

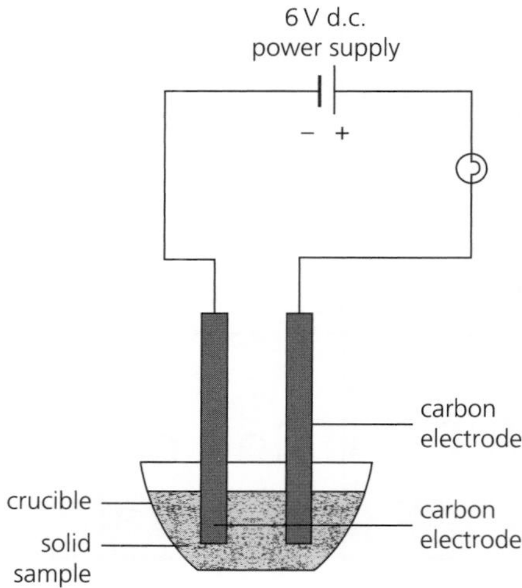

Figure 1

Method 1 – solids

1. Put on your eye protection.
2. Set up the circuit as shown in Figure 1 but do not connect the battery or switch on your power supply.
3. Before switching on ask your teacher to check your circuit.
4. Add a few spatulas of copper(II) sulfate to the crucible.
5. Switch on the power supply (set to 6 V) or connect the battery. Hold the electrodes in contact with the solid in the crucible.
6. Record your observations in Table 1 in the Observations section.
7. Repeat steps 2 to 6 with magnesium sulfate, sugar, a sample of wax and a strip of polythene.

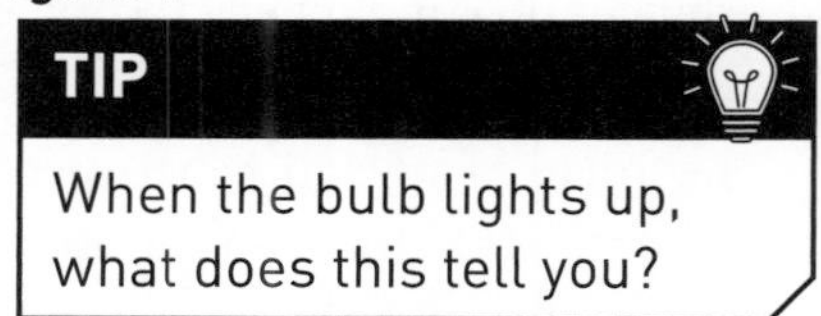

TIP

When the bulb lights up, what does this tell you?

Method 2 – solutions

In this experiment polythene and candle wax are not used. This is because polythene and candle wax do not dissolve in water.

1. Put on your eye protection.
2. Set up the circuit as shown in Figure 2 but do not connect the battery or switch on your power supply.
3. Quarter fill the beaker with copper(II) sulfate solution and check the power supply (if using) is set to 6 V.
4. Before switching on ask your teacher to check your circuit.

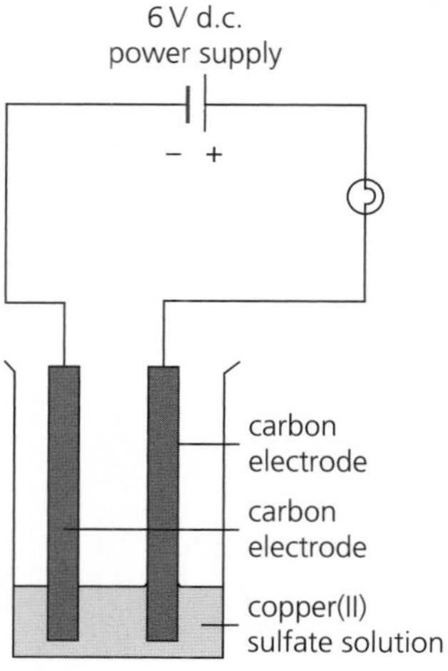

Figure 2

5 Switch on the power supply or connect the battery.

6 Record your observations in Table 2 in the Observations section.

7 Switch off the electricity. Pour away the copper(II) sulfate solution and rinse the electrodes with distilled/deionised water.

8 Repeat steps 2 to 7 with magnesium sulfate solution and sugar solutions.

Observations

Table 1

Solid	Does the bulb light?	Ionic or covalent?
Magnesium sulfate		
Copper(II) sulfate		
Sugar		
Polythene		
Candle wax		

Does this experiment allow you to distinguish between ionic and covalent substances? Explain your answer.

....

....

....

Table 2

Solution	Does the bulb light?	Observation at negative electrode	Observation at positive electrode	Conclusion
Magnesium sulfate				
Copper(II) sulfate				
Sugar				

Conclusions

1 Complete the sentences:

None of the .... conducted electricity.

You cannot identify if these are ionic or covalent substances by checking their ....conductivity.

When an electric current is passed through a solution and the bulb ....

then the dissolved substance is .... If the bulb does not

.... then the dissolved substance is ....

Therefore the covalent substances are ....and .... The ionic substances are

.... and ....

2 Explain why solid copper(II) sulfate does not conduct electricity but a solution of copper(II) sulfate does.

....

....

....

3 Explain why pure (distilled water) does not conduct electricity.

....

....

Evaluation

1 How could the methods be improved to give a more quantitative comparison of conductivity?

....

....

2 Describe other tests you could carry out to check your answers to whether the substance was ionic or covalent.

....

....

GOING FURTHER

Suggest why it is difficult to test the electrical conductivity of simple molecular substances such as oxygen and nitrogen.

....

....

....

4 Stoichiometry – chemical calculations

4.1 Determination of the empirical formula of magnesium oxide

Magnesium is a reactive metal which will, when heated, react with oxygen gas in air to form solid magnesium oxide. When the reaction occurs, an intense white light is observed.

magnesium + oxygen ⟶ magnesium oxide

KEY TERMS 

empirical formula
mole

Even at room temperature, magnesium metal reacts slowly with oxygen to produce the oxide, so before the experiment starts it is necessary to remove the oxide layer on the surface of the magnesium metal using some emery paper.

The formula of magnesium oxide can be found from the moles of oxygen gas which will react with one mole of magnesium metal, using an empirical formula calculation.

Aim

To find, by experiment, the empirical formula of the compound magnesium oxide by burning a known mass of magnesium metal in air.

Apparatus and chemicals

- Eye protection
- Crucible and lid
- Bunsen burner
- Heat-resistant mat
- Tripod
- Pipe-clay triangle
- Emery paper
- Top pan balance (minimum resolution 0.01 g)
- Tongs that meet
- 10 cm of magnesium ribbon

SAFETY GUIDANCE 

- Eye protection must be worn.
- Make sure that you do not look directly at the intense white light produced when the magnesium burns because it can damage your eyes.
- Take care to avoid burns by using tongs when lifting the crucible lid, and allow the crucible to cool before moving it.
- Magnesium ribbon – low hazard
- Magnesium oxide – low hazard

Method

Throughout the practical the student should wear eye protection.

1 Clean a 15 cm piece of magnesium ribbon by pulling the ribbon through a folded up piece of emery paper a few times – it should become shinier as you remove the magnesium oxide and other impurities from its surface.

2 Weigh the crucible and lid. Make sure that the lid you are using overlaps the edges of the crucible. Record the mass in the Observations section.

3 Coil up your magnesium ribbon so that it fits onto the bottom of the crucible.

4 Weigh the crucible, lid and magnesium, again recording the mass in the Observations section.

5 Place the crucible containing the magnesium, and the lid, onto a pipe-clay triangle on a tripod above a Bunsen burner.

6 Using a colourless Bunsen flame, heat the crucible strongly for about 15 minutes. Every 5 minutes lift off the crucible lid, with the tongs, for 2–3 seconds. During this time the magnesium metal will have reacted with the oxygen in the air. If it has not, keep heating for another 5 minutes.

TIP 

Usually a Bunsen burner, when lit, has a closed valve and an orange flame is seen. By opening the valve the flame becomes colourless. Throughout this experiment you must heat the crucible with a colourless flame as less soot is produced due to more oxygen gas being allowed in the burning gas mixture.

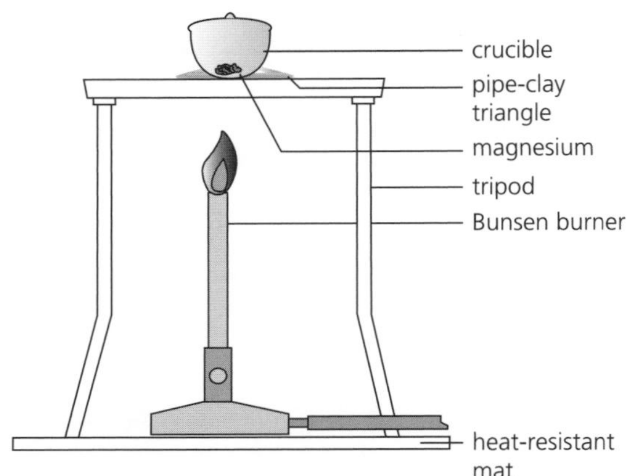

Figure 1

7 After the reaction has occurred, allow the crucible, magnesium oxide and lid to cool.

TIP 

Allow the crucible to cool before you record its mass on the balance. This will protect the balance.

8 Reweigh the crucible, magnesium oxide and lid on the balance and record its mass in the Observations section.

9 Re-heat the crucible for another 5 minutes, raising the lid once for a few seconds. Allow the crucible, magnesium oxide and lid to cool before re-weighing them. Record the mass in the Observations section.

10 If the mass recorded in step 8 is the same as that in step 9, then the experiment is finished. If the mass has changed, repeat step 9 until it is constant. Record the final constant mass in the Observations section.

Observations

Mass of crucible and lid = ... g

Mass of crucible, lid and magnesium ribbon = ... g

Mass of crucible, magnesium oxide and lid after 15 minutes = ... g

Final mass of crucible, magnesium oxide and lid after further heating = ... g

1 From your results, find the mass of magnesium used in the experiment.

....

2 From your results, find the mass of oxygen gas from the air which has reacted with the magnesium in your experiment.

....

3 Use the masses from points 1 and 2 to complete the following calculation to obtain the formula of magnesium oxide from your experiment.

Table 1

	Mg	**O**
Mass /g		
Moles	$\frac{}{24}$ =	$\frac{}{16}$ =
Simplest ratio of moles		

Conclusion

What is the empirical formula of magnesium oxide from your data?

....

Evaluation

1 State why the magnesium ribbon is coiled up at the start of the experiment.

....

....

2 Explain why it is important to use a colourless Bunsen flame to heat the crucible.

....

....

3 Why is a pipe-clay triangle used and not a gauze?

....

....

4 Why is it important to raise the lid during the experiment?

....

....

5 Why is it important only to raise the lid for a few seconds?

....

....

6 Why does the heating have to be repeated until there is no change in the final mass of the apparatus and its contents?

....

....

7 There are other gases in air which magnesium can react with to produce other products. How could the procedure be changed to remove this problem?

....

....

8 Describe how the procedure could be improved to increase the accuracy of the data obtained.

....

....

4.2 Determination of the volume occupied by one mole of a gas

It is known that at 25 °C and 1 atmosphere of pressure, one mole of any gas occupies a volume of 24 dm^3. This is known as the molar volume.

In this experiment we will use the reaction between a known mass of magnesium metal and excess hydrochloric acid to give hydrogen gas, which will be collected and its volume recorded.

magnesium + hydrochloric acid → magnesium chloride + hydrogen

$$Mg(s) + 2HCl(aq) \rightarrow MgCl_2(aq) + H_2(g)$$

KEY TERM

mole

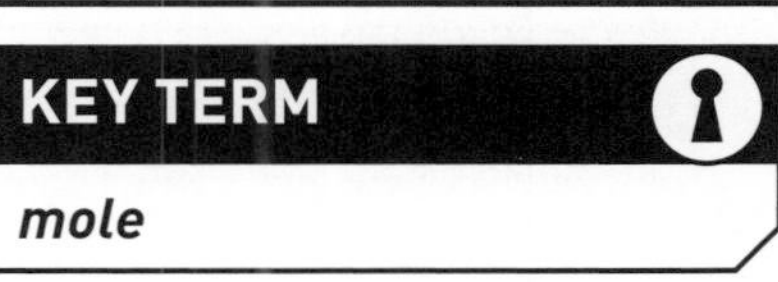

TIP

If a 50 cm^3 burette cannot be used, consider using a 50 cm^3 measuring cylinder under water, but this will not be as accurate as using an inverted burette.

Aim

To determine the volume occupied by one mole of hydrogen gas at the temperature and pressure of your chemistry laboratory.

Apparatus and chemicals

- Eye protection
- 50 cm^3 burette

- Clamp stand, boss and clamp
- Boiling tube
- Bung with single hole fitted with delivery tube
- Basin
- 10 cm^3 measuring cylinder
- Boiling-tube rack
- Top pan balance (minimum resolution 0.01 g)
- Scissors
- Emery paper
- 10 cm of magnesium ribbon
- 2 mol/dm^3 hydrochloric acid
- Thermometer

SAFETY GUIDANCE

- Eye protection must be worn.
- There should be no flames in the room. Hydrogen gas is flammable.
- Magnesium ribbon – low hazard
- Magnesium chloride solution – low hazard
- Hydrogen gas – extremely flammable

Method

Throughout the practical the student should wear eye protection.

1 Using the measuring cylinder, place 5 cm^3 of the 2 mol/dm^3 hydrochloric acid into the boiling tube and place it into the boiling-tube rack.

2 Clean the magnesium ribbon with the emery paper.

3 Weigh the magnesium ribbon using the balance. Use scissors to get the mass of the ribbon as close to 0.04 g as possible. Record the actual mass of the ribbon you use in the Observations section, to two decimal places.

4 Set up the burette up as shown in Figure 2. You will need to fill the burette with water and then, with your finger over the open end, invert it into the basin of water. You may need to open the tap on the burette to lower the water level so that it starts on the scale on the burette.

5 Take the initial burette reading, to two decimal places (or one decimal place if you are using a measuring cylinder).

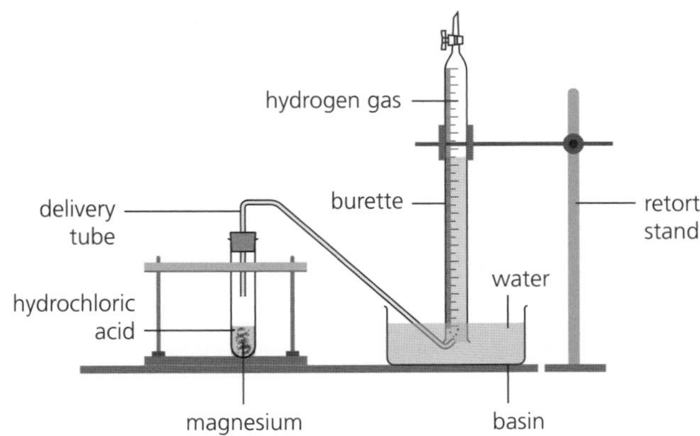

Figure 2

6 To start the reaction quickly but safely, add the 0.04 g of magnesium to the hydrochloric acid in the boiling tube and replace the bung.

TIP

It is important to be able to take a burette reading accurately. You must look at the bottom of the meniscus (the lowest part of the curve the water forms in the burette), at eye level to rule out parallax errors. In this experiment the scales will be upside down so take care when taking your readings.

TIP

Burettes can be read to two decimal places, that is to the nearest 0.05 cm^3, where the second decimal place is either a 0, if the bottom of the meniscus sits on the scale division, or a 5 if it is between the divisions. So if, for example, the meniscus was between 49.10 cm^3 and 49.20 cm^3, the reading would be 49.15 cm^3.

The measuring cylinder would only be able to be read to the nearest 0.5 cm^3.

7 After all of the magnesium ribbon has reacted, record the final volume of hydrogen gas collected in the burette.

8 Using the thermometer, record the temperature of the laboratory.

Observations

Mass of magnesium ribbon used = ... g

Volume of gas collected = ... cm^3

Temperature = ... °C

1 From your results, calculate the number of moles of magnesium you used in your experiment. Give your result to two significant figures.

> **TIP**
>
> To round an answer to two significant figures, look at the third digit. Round up if the digit is 5 or more, and round down if the digit is 4 or less.

....

....

2 Using the balanced chemical equation, work out the number of moles of hydrogen gas that could be produced from the number of moles of magnesium you have just calculated.

....

....

....

3 The number of moles of hydrogen you have just calculated is equivalent to the volume of gas collected in the burette. Determine what the volume would have been if you had produced one mole of hydrogen gas.

....

4 Convert the volume in cm^3 into dm^3. Give your result to two significant figures.

....

> **TIP**
>
> Use the unrounded value for the number of moles in this calculation. Rounding should only be done at the end.

Conclusions

1 From your results, what is the molar volume of hydrogen gas under the conditions at which the experiment was carried out?

....

2 Suggest reasons why your method produced a different result to the known molar volume of gas, which is 24 dm^3.

....

....

....

Evaluation

1 Why is the magnesium ribbon cleaned with emery paper?

....

....

2 Why is an excess of hydrochloric acid used in the experiment?

....

3 Identify two possible sources of error in the measurements in this experiment.

....

....

....

....

4 Discuss how the experiment could be improved to give more accurate results.

....

....

GOING FURTHER

The volume of a given mass of a gas changes at different temperatures and pressures. Use your research skills to find the names of the two 'laws' which explain how the volume of a gas varies with temperature and pressure.

....

4.3 Determination of the percentage yield of a chemical reaction

KEY EQUATION 

$$\% \text{ yield} = \frac{\text{actual yield}}{\text{theoretical yield}} \times 100$$

KEY TERM 

percentage yield

SAFETY GUIDANCE 

- Eye protection must be worn.
- When you put the pipette filler on, ensure that you hold the pipette close to the end the filler will be attached to. This should stop the pipette from breaking as you push on the filler.
- Sodium carbonate solution – low hazard
- Calcium nitrate solution – irritant
- Calcium carbonate solid – irritant
- Sodium nitrate solution – low hazard

In this experiment, you will react sodium carbonate with excess calcium nitrate to produce calcium carbonate, which is insoluble in water, a precipitate. You will measure the mass of the precipitate.

sodium carbonate + calcium nitrate → calcium carbonate + sodium nitrate

$$Na_2CO_3(aq) + Ca(NO_3)_2(aq) \rightarrow CaCO_3(s) + 2NaNO_3(aq)$$

Using this balanced chemical equation, you will work out the theoretical yield of calcium carbonate which can be formed by this reaction. This is the maximum amount of calcium carbonate which could be formed from the amounts of chemicals we start with and if the reaction is carried out perfectly.

By carrying out the experiment we obtain the actual yield, which is often lower than the theoretical yield.

The percentage yield of a chemical reaction shows how much product is obtained as a percentage of the maximum possible mass, as predicted by the balanced equation for the reaction.

Aim

To carry out a precipitation reaction and then to determine the mass of the precipitate formed. The percentage yield for the reaction will then be determined.

Apparatus and chemicals

- Eye protection
- 25 cm^3 pipette and filler
- 50 cm^3 measuring cylinder
- 150 cm^3 beaker
- 250 cm^3 conical flask
- Glass rod
- Filter paper
- Filter funnel
- Accurate balance
- 1 mol/dm^3 sodium carbonate solution
- 1 mol/dm^3 calcium nitrate solution
- Wash bottle with distilled water

Method

Throughout the practical the student should wear eye protection.

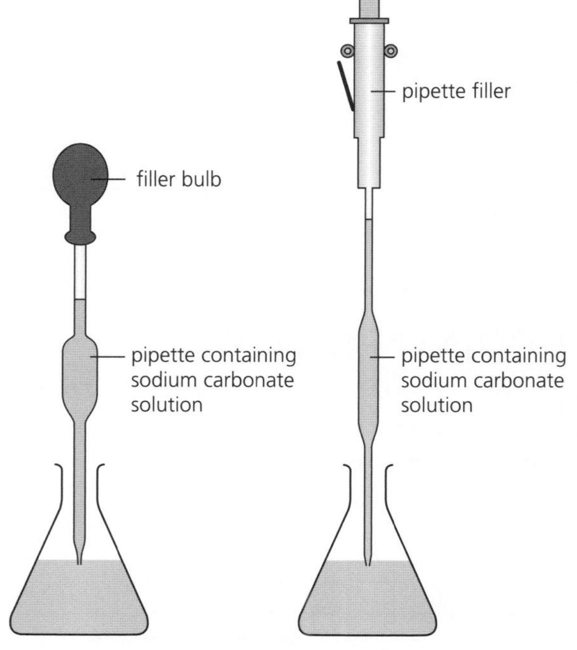

Figure 3 Two types of pipette filler

1 Using the pipette, with a filler, place 25.0 cm^3 of 1 mol/dm^3 sodium carbonate solution into the beaker.
2 Using the measuring cylinder, pour about 50 cm^3 of 1 mol/dm^3 calcium nitrate into the beaker.
3 Using the glass rod, stir the contents of the beaker gently for 2 minutes.
4 Use the distilled water to ensure that all the solid is removed from the glass rod into the beaker.
5 Use the balance to weigh a piece of the filter paper, and record the mass in the results section.
6 Put the filter paper into the funnel, place the funnel in the top of the conical flask and filter the contents of the beaker. Use the distilled water to ensure that all of the solid has been removed from the beaker and is tipped onto the filter paper.
7 The calcium carbonate and filter paper now need to dry. This can be done by placing the filter paper over a radiator, using a drying oven or simply by leaving it for a few days.
8 When dry, find the mass of the filter paper and the calcium carbonate, and record the mass in the Observations section.

Observations

Mass of filter paper = ... g

Mass of dry filter paper and calcium carbonate = ... g

1 From your results, calculate the mass of calcium carbonate produced in the experiment. This is the actual yield.

....

We now need to determine the theoretical yield using the amount of sodium carbonate used and the balanced equation.

2 Calculate the number of moles of sodium carbonate in 25 cm^3 of 1 mol/dm^3 solution.

....

....

3 Using the equation, determine the number of moles of calcium carbonate that should be produced.

....

....

4 Using the number of moles of calcium carbonate which should have been produced, calculate the mass that should have been obtained. This is the theoretical yield. (A_r: Ca = 40, C = 12, O = 16)

....

....

5 Now work out the percentage yield for the experiment. Give your answer to three significant figures.

....

....

Conclusion

Was the percentage yield good or bad?

....

....

Evaluation

1 Identify two possible reasons the yield is lower than you would expect in this experiment.

Evaluate which of the two reasons you have given had the most effect.

....

....

2 State why an excess of calcium nitrate solution was used.

....

3 Explain why a pipette was used to measure out the sodium carbonate solution, but a measuring cylinder was used for the calcium nitrate solution.

....

....

....

4 Why was it necessary to dry the filter paper and calcium carbonate before weighing them?

....

....

....

5 Why is the calcium carbonate not simply removed from the filter paper to find the mass produced?

....

....

6 Write a balanced ionic equation for the reaction you have just carried out.

....

....

GOING FURTHER

The atom economy of a reaction is a measure of the amount of the chemical reactants that end up in the useful product from a reaction. It can be determined using:

$$\text{atom economy} = \frac{\text{mass of atoms in the desired product}}{\text{mass of atoms in the reactants}} \times 100$$

Hydrogen gas, a raw material in the Haber process, can be made in two ways:

$CH_4 + H_2O \rightarrow CO + 3H_2$

$CO + H_2O \rightarrow CO_2 + H_2$

State which of these two methods of producing hydrogen gas has the highest atom economy, and why it would be important for industry to use the method with the highest atom economy.

....

....

....

....

5 Electrochemistry

5.1 Electrolysis of acidified water (Teacher demonstration)

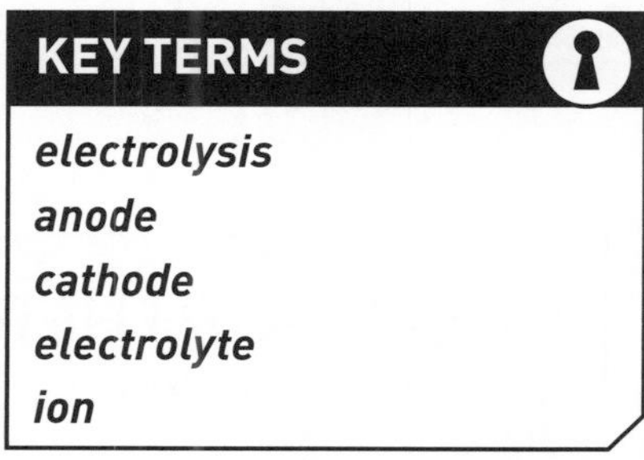

Several industrial processes involve the electrolysis of aqueous solutions. Pure water is a covalent compound and as such is a very poor conductor of electricity. It is known that a *small* proportion of molecules in liquid water break down to form hydrogen ions, $H^+(aq)$, and hydroxide ions, $OH^-(aq)$. However, there are still not enough ions for water to be a good conductor.

However, if some sulfuric acid is added to the water, then the sulfuric acid provides ions to the solution and this allows an electric current to pass through the water. The water then decomposes (breaks down to form elements). The two gaseous elements produced are, of course, hydrogen and oxygen!

The standard apparatus used to carry out this experiment is the Hofmann voltameter (Figure 1).

An alternative piece of apparatus can be constructed using two burettes over the electrodes so that the gases can be kept separate throughout the demonstration.

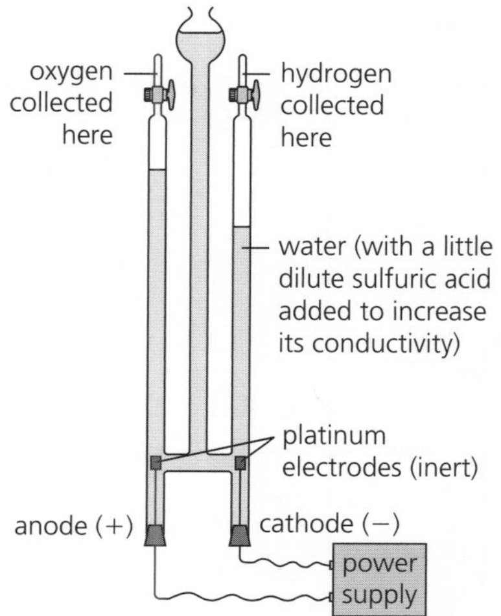

Figure 1

Aim

To electrolyse acidified water and confirm the formula for water is H_2O.

Apparatus and chemicals

- Eye protection
- 400 cm^3 beaker
- Hofmann voltameter
- 2 × test tubes
- 2 × wooden spills
- Low voltage, d.c. power supply
- Deionised water
- 1 mol/dm^3 sulfuric acid

SAFETY GUIDANCE

- Eye protection must be worn.
- Sulfuric acid (1 mol/dm^3) – irritant
- Hydrogen – flammable. Make sure there are no flames in the room.

Method

The teacher or demonstrator and students should wear eye protection throughout the demonstration experiment.

1. Mix some deionised water ($\frac{2}{3}$ by volume) with dilute sulfuric acid ($\frac{1}{3}$ by volume) in a 400 cm^3 beaker.
2. Pour this solution into the apparatus through the reservoir (a thistle-headed centre tube) until both of the outside tubes are full with the taps open. Then close them.
3. Connect the power supply and switch on. Collect the gases produced in the anode tube and cathode tube for approximately 10 minutes.
4. Record the approximate ratio of the volumes in the cathode tube : the anode tube.

5 Place test tubes over these tubes and collect the gases in the test tubes.

6 Test the gas from the anode tube with a glowing splint and test the gas collected from the cathode tube with a lighted splint.

Observations

Volume of gas collected at the:

cathode = ...

anode = ...

Approximate ratio of gases produced, cathode : anode is ....

Effect of glowing splint with anode tube gas: ....

The gas is ....

Effect of lighted splint with cathode tube gas: ....

The gas is ....

Do the results fit in with the known formula for water?

Explain your answer.

....

....

Conclusion

Complete the sentences.

Pure water is a very bad .... of electricity. This is because it has very few .... present in the liquid. If dilute .... acid is added to water, ... are added and this makes the solution a .... one. An electric current passes through this solution and .... takes place. Hydrogen gas is produced at the .... and oxygen gas is produced at the .... The ratio of the volumes of the gases produced is .... hydrogen to oxygen. This fits in with the formula for water being ....

Evaluation

1 Why is deionised water used in this experiment?

....

2 Outline why rinsing the apparatus and electrodes with distilled water before the experiment might be a good idea.

....

....

3 Why was the power supply connected up *after* the apparatus is filled by the solution?

....

....

4 Write word and balanced ionic half-equations for the processes that take place at the anode and cathode.

....

....

....

....

GOING FURTHER

Polypyrrole is a plastic that can conduct electricity. Which parts of the apparatus you have seen used today could be replaced by this plastic?

....

....

Would you expect different results to this experiment? Explain your answer.

....

....

5.2 Electrolysis of brine (Teacher demonstration)

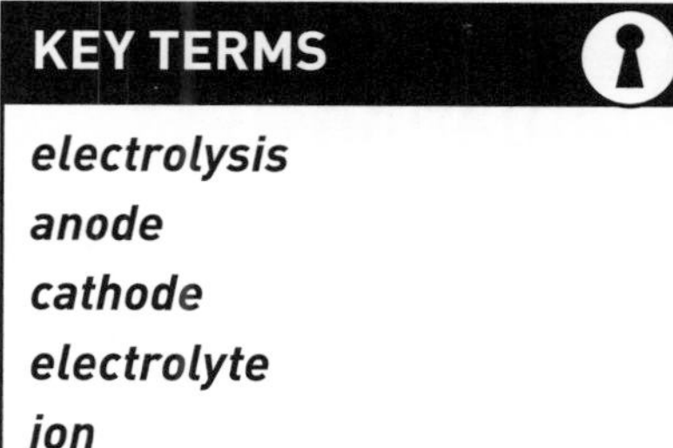

The electrolysis of saturated sodium chloride solution (brine) is the basis of a major industry – the chlor-alkali industry. The electrolytic process is a very expensive one, requiring vast amounts of electricity. The process is economic only because all three products, hydrogen, chlorine and sodium hydroxide, have a large number of uses.

The standard apparatus used to carry out this experiment is the Hofmann voltameter (see Figure 1 on page 46).

If a Hofmann voltameter is not available, an alternative (which still allows the gases to be kept separate) is to collect the gaseous products in two burettes inverted over the electrodes.

Aim

To electrolyse concentrated sodium chloride solution (brine) and identify the products.

SAFETY GUIDANCE 

- Eye protection must be worn.
- Saturated sodium chloride solution (brine) – low hazard
- Universal indicator solution – low hazard
- Chlorine – toxic. Chlorine, even in a small quantity, can induce an asthma attack and so it should not be inhaled. Switch the power off as soon as a small amount of gas has been collected to reduce the amount of chlorine released into the room.
- Hydrogen – extremely flammable. Make sure there are no flames in the room.
- Sodium hydroxide solution irritant, corrosive

Apparatus and chemicals

- Eye protection
- Hofmann voltameter
- 2 × test tubes
- 2 × wooden spills/splints
- Low voltage, d.c. power supply
- Saturated sodium chloride solution
- Universal indicator solution
- Access to fume cupboard

Method

The teacher or demonstrator and students should wear eye protection throughout the demonstration experiment. It should take place in a fume cupboard or a well-ventilated laboratory.

1 Students should record the results of any experiments with collected gas and colour changes during the demonstration.

2 Add a little universal indicator solution to the saturated sodium chloride solution.

3 Pour this solution into the apparatus through the reservoir (a thistle-headed centre tube) until both of the outside tubes are full.

4 Connect the power supply and switch on. Collect any gases produced in the anode tube and cathode tube for approximately 10 minutes.

5 Note down in the Observations section any colour changes that have taken place.

6 Place a test tube over the cathode tube and collect the gas produced in it.

7 Test the gas collected from the cathode tube with a lighted splint.

Observations

Make predictions related to the experiment you have seen demonstrated.

What gases may be produced and what is the substance left in solution?

....

....

Colour changes of the solution in the anode tube:

Initially ....

During the experiment ....

Colour changes of the solution in the cathode tube:

Initially ....

During the experiment ....

Effect of lighted splint with cathode tube gas:

....

The gas is ....

Do the results fit in with the expectation of the electrolysis? ....

Explain your answer.

....

....

....

Conclusion

1 Complete the sentences.

Brine is a very good .... of electricity. This is because it has a lot of .... present in the liquid. This makes the solution an .... one. An electric current passes through this solution and .... takes place.

Hydrogen gas is produced at the .... and .... gas is produced at the .... is left in solution.

2 Write word and balanced ionic half-equations for the processes that take place at the anode and cathode.

....

....

....

....

....

....

Evaluation

1 Outline why rinsing the apparatus and electrodes with brine before the experiment might be a good idea.

....

....

2 Why is the experiment carried out in a fume cupboard or well-ventilated laboratory?

....

3 Why is universal indicator solution used in this experiment?

....

....

4 Why is the power supply connected up after the apparatus is filled by the solution?

....

....

GOING FURTHER

How do you think the method you used today to electrolyse brine could be upgraded to a large scale?

....

....

....

....

....

....

6 Chemical energetics

6.1 Calculating the enthalpy change for the combustion of methanol and ethanol

The enthalpy change for the combustion of any fuel is the energy given off when the fuel is completely burned in oxygen. The enthalpy change for a particular reaction depends on how much fuel is burned, so it is more useful to find the enthalpy change per gram of fuel, or per mole of fuel.

KEY TERMS 

enthalpy change
combustion

In this experiment, a known mass of water will be heated by a fuel in a spirit burner. The temperature rise of the water and the mass of fuel burned can then be used to find the energy transferred from the fuel to the water using:

energy transferred (J) = mass of water used (g) × 4.2 (J/g °C) × rise in temperature (°C)

where the specific heat capacity of water is 4.2 J/g °C.

The density of water is 1.0 g/cm³.

Aim

To determine the enthalpy change of combustion of methanol and ethanol.

SAFETY GUIDANCE 

- Eye protection must be worn.
- Methanol - flammable, toxic.
- Ethanol - flammable.
- The copper calorimeter will be very hot at the end of the first experiment.
- Take care when using spirit burners. Depending on the length of wick sticking out of the top of the burner, the flame size can get very large.

Apparatus and chemicals

- Eye protection
- Metal calorimeter
- Glass rod
- Spirit burner containing methanol and lid
- Spirit burner containing ethanol and lid
- Clamp stand, boss and clamp
- 100 cm³ measuring cylinder
- Thermometer
- Accurate balance

Method

Throughout the practical the student should wear eye protection.

1 Using a measuring cylinder, put 150 cm³ (or a known volume) of cold water into the metal calorimeter. Take the temperature of the water using a thermometer, and record it in the Observations section.

2 Clamp the calorimeter over a spirit burner containing methanol. Ensure that the height of the calorimeter is set so that the flame from the burner will meet the bottom of the calorimeter (Figure 1).

3 Weigh the spirit burner with the lid on. Record your results in the Observations section.

4 Replace the burner under the calorimeter, remove the lid and light the wick. Use the thermometer to stir the water all the time it is being heated. Continue heating until the temperature has risen by between 25 and 30 °C.

5 Extinguish the burner by blowing out the flame and replace the lid.

6 Keep stirring the water and record the highest temperature reached in the Observations section.

7 Weigh the spirit burner and lid again.

8 Repeat steps 1–7 using a spirit burner containing ethanol. Put new cold water into the calorimeter and ensure that the flame size is the same as in the first experiment.

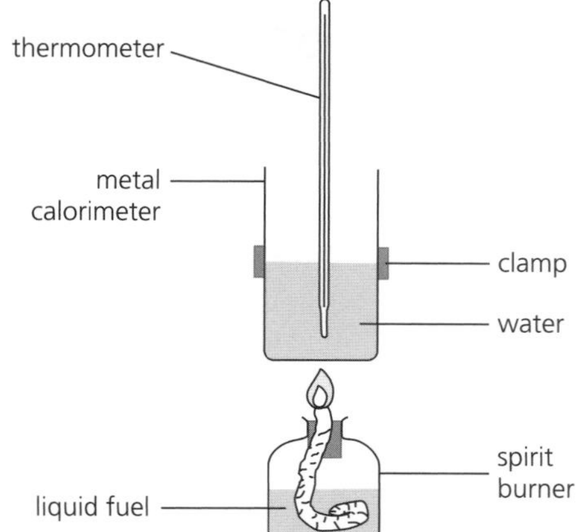

Figure 1

TIP

Stir the water in the calorimeter whilst the water is being heated to ensure that the temperature is the same throughout the water.

Observations

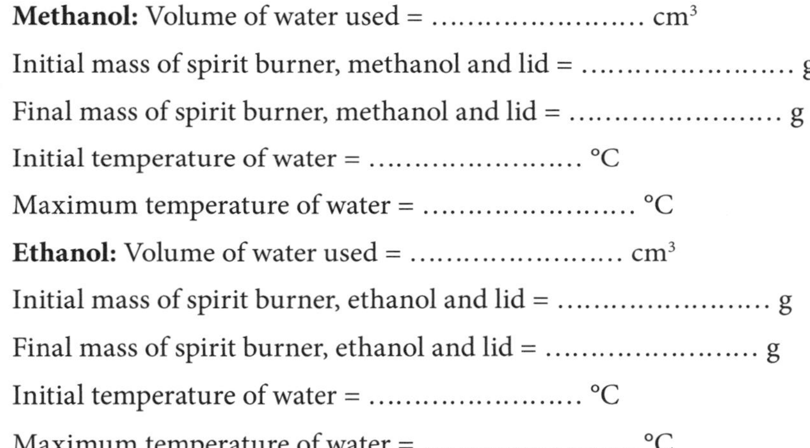

Methanol: Volume of water used = ... cm^3

Initial mass of spirit burner, methanol and lid = ... g

Final mass of spirit burner, methanol and lid = ... g

Initial temperature of water = ... °C

Maximum temperature of water = ... °C

Ethanol: Volume of water used = ... cm^3

Initial mass of spirit burner, ethanol and lid = ... g

Final mass of spirit burner, ethanol and lid = ... g

Initial temperature of water = ... °C

Maximum temperature of water = ... °C

TIP

The calorimeter will be very hot at the end of the first experiment so handle it carefully. It will also be coated with soot so this should be washed off between experiments.

Use the table to determine the enthalpy change of combustion for methanol and ethanol.

	Methanol	**Ethanol**	
1 Find the temperature rise in the experiment / °C			
2 Energy transferred / J (mass of water × 4.2 × temp. rise)			
3 Mass of fuel burned / g			
4 Moles of fuel burned / moles (mass/M_r)	$\frac{\square}{32}$ =	$\frac{\square}{46}$ =	
5 Energy transferred by 1 mole / J/mol (energy transferred / moles burned)			
6 Enthalpy change of combustion / kJ/mol (energy transferred by 1 mole/1000) =	$\frac{\square}{1000}$ =	$\frac{\square}{1000}$ =	

Conclusions

The data-book value for the enthalpy change of combustion of methanol is –726 kJ/mol and for ethanol it is –1371 kJ/mol.

Enthalpy changes depend on the temperature and pressure at which the reactions occur. The data-book values are given for a set of standard conditions: at 25 °C (298 K) and 1 atmosphere of pressure, and for conditions in which complete combustion occurs.

1 What does the '–' indicate in front of these numbers?

....

....

Your values should have a '–' sign inserted at the front.

2 Suggest a reason why the values you have obtained for the enthalpy change of combustion are so small compared to the actual values.

....

....

....

....

....

....

Evaluation

1 Why is a metal calorimeter used instead of a glass beaker?

....

....

2 What is the purpose of the lid on the spirit burner?

....

....

....

3 At the end of the experiment what can you see has happened to the bottom of the copper calorimeter? What is this caused by?

....

....

4 As you have seen from your results, in comparison with the actual values, the experimental values are not accurate (close to the true values). Evaluate the procedure used and make a list of the major sources of error.

....

....

....

....

5 Suggest some ways in which the experiment could be improved.

....

....

....

GOING FURTHER

Use your research skills to find out how the data-book values are obtained.

....

....

6.2 Determination of the enthalpy change of a displacement reaction

Iron is a more reactive metal than copper. So when iron metal is added to copper(II) sulfate solution, a displacement reaction occurs.

iron + copper(II) sulfate → iron(II) sulfate + copper

$Fe(s) + CuSO_4(aq) \rightarrow FeSO_4(aq) + Cu(s)$

By measuring the temperature rise which occurs when the iron and copper(II) sulfate are reacted together, it is possible to classify the reaction as exothermic or endothermic.

It is also possible to work out the enthalpy change for the reaction.

KEY TERMS 

exothermic reaction
endothermic reaction

enthalpy change

Aim

To determine whether a displacement reaction is endothermic or exothermic.

To determine the enthalpy change when the displacement reaction between iron metal and copper(II) sulfate occurs.

SAFETY GUIDANCE 

- Eye protection must be worn.
- When you are putting the pipette filler onto the pipette, make sure that you hold the pipette close to the end the filler will be attached to. This should stop the pipette from breaking as you push on the filler.
- Copper(II) sulfate (1 mol/dm³) – corrosive, irritant and hazardous to the aquatic environment. Wear gloves.
- Iron(II) sulfate solution – harmful

Apparatus and chemicals

- Eye protection
- 250 cm³ beaker
- Polystyrene cup
- 25 cm³ pipette and filler
- Glass rod
- Thermometer
- Stopwatch
- Accurate balance
- 2 g of iron filings
- 1 mol/dm³ copper(II) sulfate solution – harmful/corrosive

TIP 

Be careful not to poke a hole through the polystyrene cup with the thermometer.

Method

Throughout the practical the student should wear eye protection.

1. Using the pipette and filler, place 25.0 cm³ of the 1 mol/dm³ copper(II) sulfate solution into the polystyrene cup.
2. Using the balance, weigh out about 2 g of iron filings. This is an excess.
3. Place the polystyrene cup into the glass beaker to keep it stable.
4. Put the thermometer in the copper(II) sulfate solution and record the temperature in the Observations table. Repeat every 30 seconds, recording the results in the table in the Observations section.
5. After 2 minutes, add all of the iron filings to the copper(II) sulfate solution. Continue to record the temperature of the reaction mixture every 30 seconds until 12 minutes have passed. Enter the readings, in the table in the Observations section. Stir the mixture with the thermometer to ensure that the reactants are mixed well.

Observations

Time / minutes	Temperature / °C
0	
0.5	
1.0	
1.5	
2.0	
2.5	
3.0	
3.5	
4.0	
4.5	
5.0	
5.5	
6.0	

Time / minutes	Temperature / °C
6.5	
7.0	
7.5	
8.0	
8.5	
9.0	
9.5	
10.0	
10.5	
11.0	
11.5	
12.0	

Plot the results you have obtained on the graph paper. Do not, at this point, draw any curves or lines.

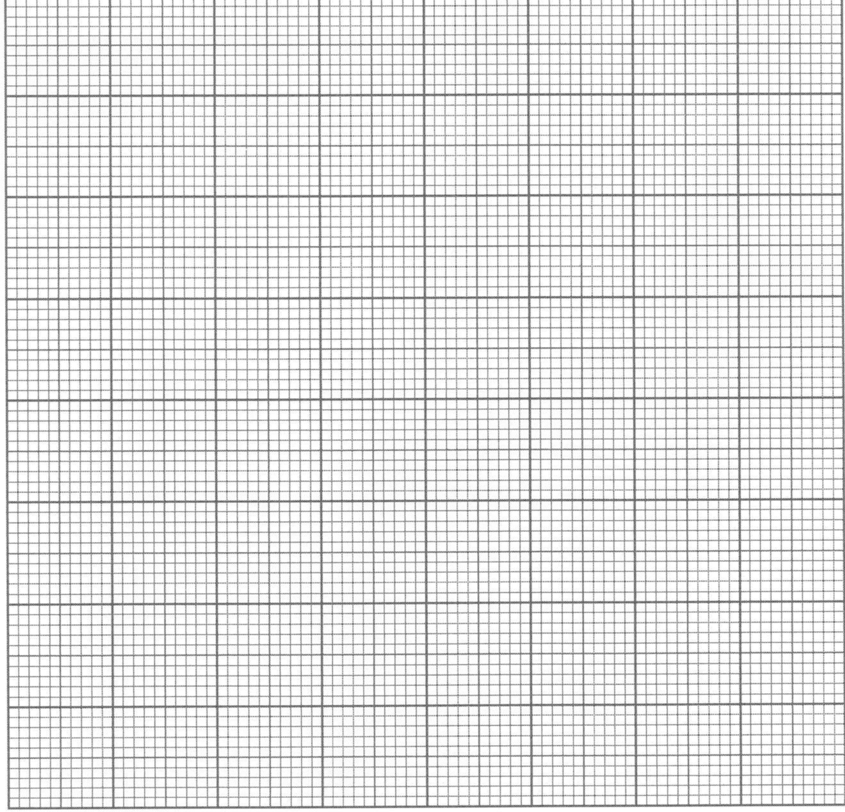

1 Is this reaction exothermic or endothermic?

....

You will notice, from the shape of your graph, that the maximum recorded temperature does not occur exactly when the iron filings have been added to the copper(II) sulfate. To work out the maximum temperature rise that would have occurred if the reaction occurred instantaneously, follow Figure 2.

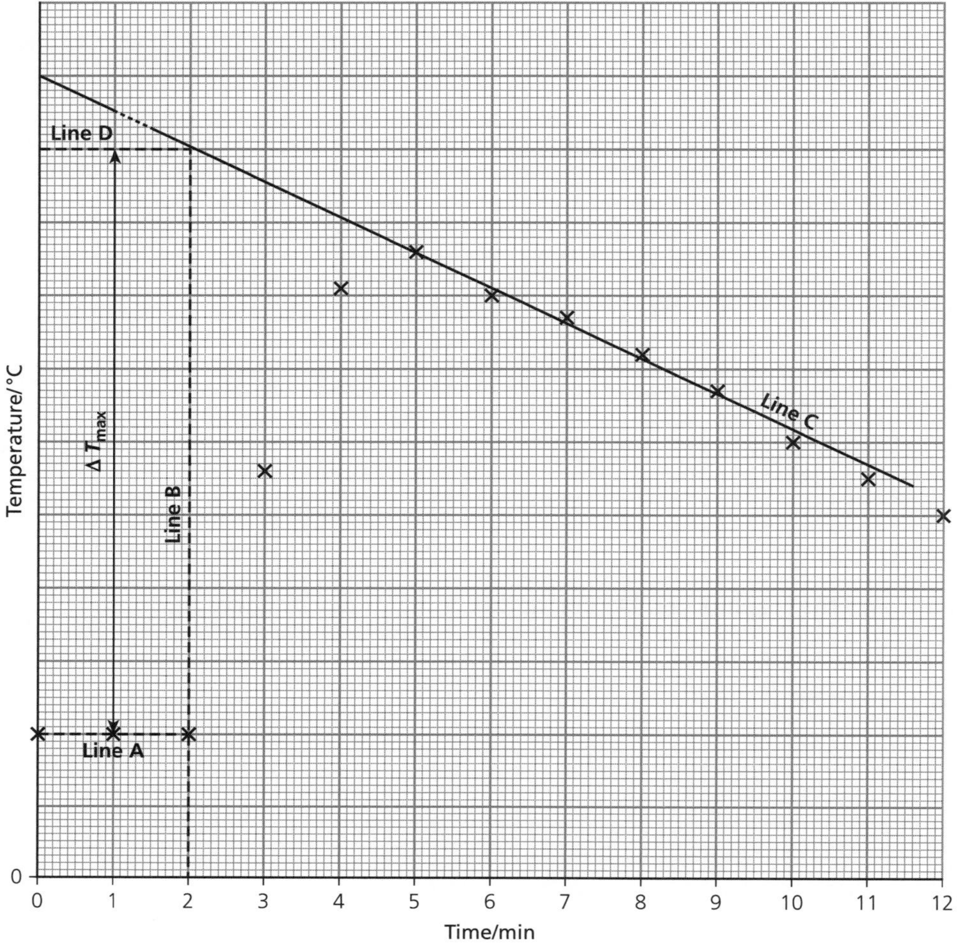

Figure 2

2 Use a ruler to draw lines, A, B, C and D on your own graph, as shown in Figure 2.

3 The theoretical maximum temperature change which could have occurred, had all the copper(II) sulfate reacted with the iron when it was added, is shown as ΔT_{max} on the diagram. Determine what ΔT_{max} would have been for your results.

....

4 Find the energy transferred from the reaction to the water using:

energy transferred (J) = mass of solution (g) × 4.2 (J/g °C) × rise in temperature (°C)

where the specific heat capacity of water is 4.2 J/g °C. Assume the density of copper(II) sulfate solution is the same as water, $1.0\,g/cm^3$. Give your answer to three significant figures.

....

5 This is the amount of energy transferred when 0.025 moles of copper(II) sulfate is used. What would have been the energy transferred if you had started with 1 mole of copper(II) sulfate?

....

....

6 For this reaction what is the enthalpy change, ΔH, to three significant figures?

....

TIP

An exothermic reaction has a *negative* ΔH: $\Delta H < 0$, and an endothermic reaction has a *positive* ΔH: $\Delta H > 0$.

Conclusions

1 The accepted value for the enthalpy change of displacement you have just carried out is –152 kJ/mol. Comment on the enthalpy change you determined from your experiment.

....

....

2 The enthalpy change of displacement for the reaction between copper(II) sulfate solution and zinc metal is –219 kJ/mol. Can you explain why one reaction is more exothermic than the other?

TIP

Your value should also have a '–' sign inserted at the front, so both reactions have negative enthalpy changes. You should compare the *size* of the enthalpy changes.

....

....

....

....

3 What can you predict about the size of the enthalpy change for the displacement reaction between copper(II) sulfate and tin metal, in comparison to the energy changes for iron and zinc?

....

....

Evaluation

1 Explain why a polystyrene cup is used instead of a glass beaker.

....

....

2 State why a pipette was used to measure the copper(II) sulfate and not a measuring cylinder.

....

....

....

....

3 Why was an excess of iron filings used?

....

4 When you calculated the enthalpy change of the reaction, what assumptions did you make?

....

....

5 Write down at least one way in which this experiment could have been improved and state why this would have improved the experiment.

....

....

....

....

7 Chemical reactions

7.1 How does changing surface area affect the rate of a reaction?

To show the effect of changing the surface area of a solid reactant on the reaction rate, we will use the reaction between limestone (calcium carbonate, $CaCO_3$) and hydrochloric acid.

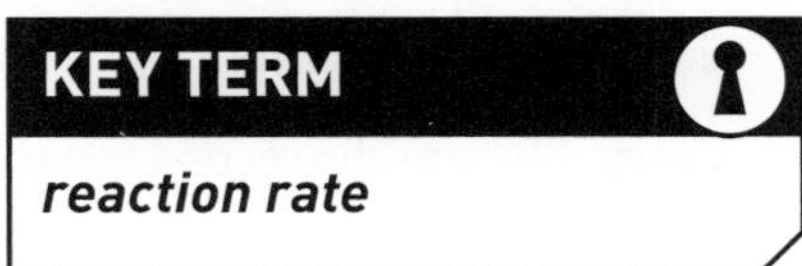

calcium carbonate + hydrochloric acid ⟶ calcium chloride + water + carbon dioxide

$$CaCO_3(s) + 2HCl(aq) \rightarrow CaCl_2(aq) + H_2O(l) + CO_2(g)$$

When this reaction occurs, carbon dioxide gas is produced. The rate of the reaction can be followed in a number of ways: because a gas is produced in this reaction one way is to look at the loss in mass of the apparatus, as the gas escapes, against time.

Aim

To find out how changing the surface area of a specific mass of solid reactant affects the rate of a chemical reaction.

Apparatus and chemicals

- Eye protection
- Top pan balance (minimum resolution 0.1 g)
- 2 × 250 cm³ conical flasks
- 100 cm³ measuring cylinder
- Cotton wool
- Stopwatch
- 2 mol/dm³ hydrochloric acid solution
- 10 g of large limestone chips
- 10 g of smaller limestone chips

SAFETY GUIDANCE 

- Eye protection must be worn.
- Hydrochloric acid (2 mol/dm³) – irritant

Method

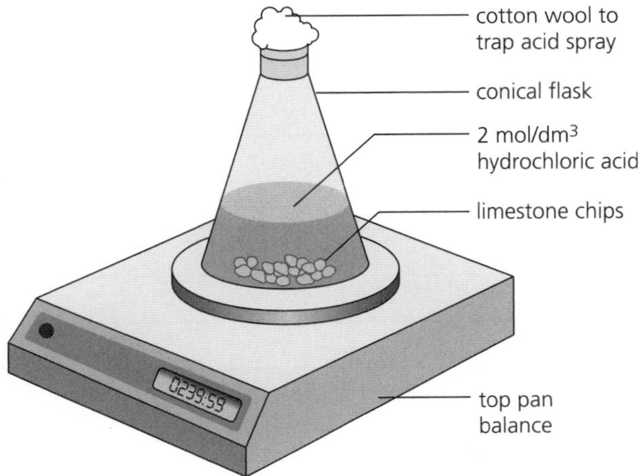

Figure 1

Throughout the practical the student should wear eye protection.

1 Place the 10 g of large limestone chips carefully into a conical flask.

2 Take a piece of cotton wool and check that it fits into the neck of the flask without falling in, and then remove it.

3 Using the measuring cylinder, measure out exactly 100 cm^3 of the 2 mol/dm^3 hydrochloric acid solution.

4 Start the stopwatch and record the mass shown on the balance as soon as you add the acid to the flask. Put the cotton wool back in the flask as quickly as you can.

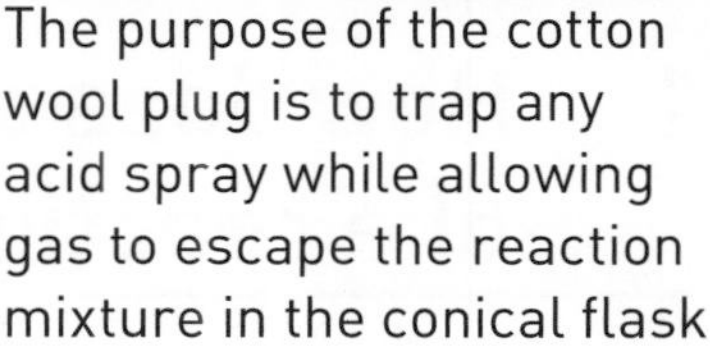

TIP

The purpose of the cotton wool plug is to trap any acid spray while allowing gas to escape the reaction mixture in the conical flask.

5 Every 30 seconds, record the mass in the table below.

6 Continue taking readings for 15 minutes, or until the mass stops changing. Remove the flask from the balance.

7 Repeat steps 1–6 for the 10 g of smaller limestone chips.

Observations

1 Complete this table as you obtain the results.

2 You will need to work out the total loss in mass, based on the initial mass, after each result is recorded.

Table 1

	10 g of large limestone chips		10 g of smaller limestone chips	
Time / min	**Mass / g**	**Loss in mass / g**	**Mass / g**	**Loss in mass / g**
0		–		–
0.5				
1.0				
1.5				
2.0				
2.5				
3.0				
3.5				
4.0				
4.5				
5.0				
6.0				
7.0				
8.0				
9.0				
10.0				
11.0				
12.0				
13.0				
14.0				
15.0				

3 Plot a graph of loss in mass (y-axis) against time (x-axis) for both experiments. Draw a smooth curved line of best fit. Use the same axes for the two curves.

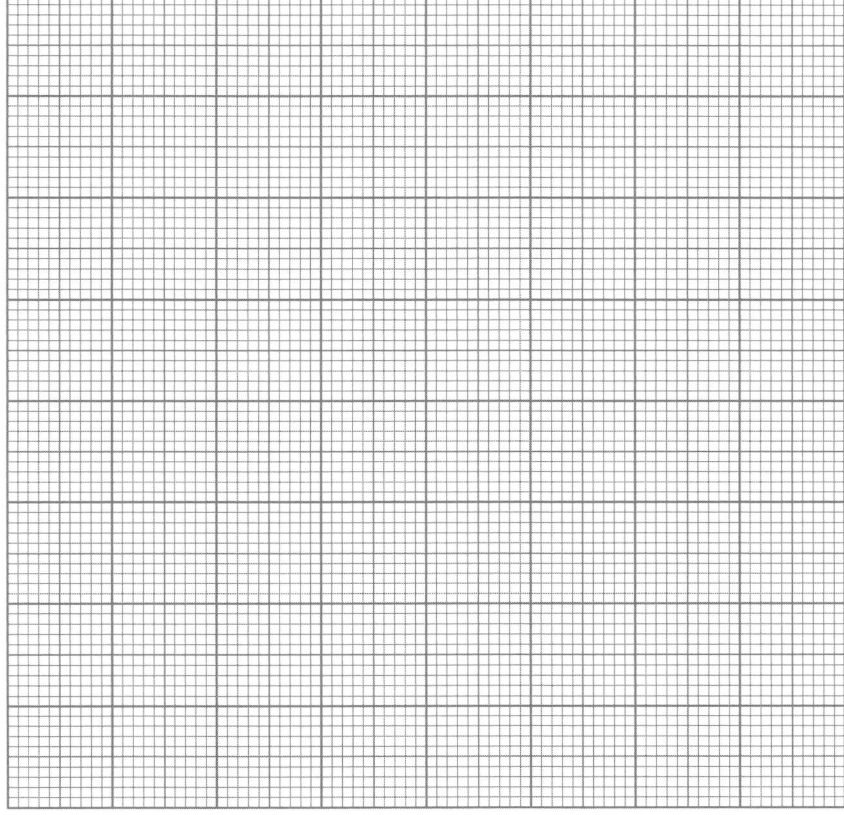

Conclusions

1 Which of the experiments has the fastest initial rate of reaction?

....

2 How can you tell this from the graph?

....

3 Use ideas about particles to explain your answer.

....

....

....

....

....

....

4 State two reasons why cotton wool is used in the neck of the conical flask.

....

....

Evaluation

1 State the variables that you kept the same for the two experiments.

....

....

2 What other factor, or factors, should be kept constant to show the effect of surface area on the rate of reaction? Were you able to control these?

....

....

3 Write down at least two ways in which this experiment could have been improved and state why they would have improved the experiment.

....

....

....

....

4 Suggest one other way the rate of reaction could have been followed, other than by loss in mass.

....

....

7.2 How does changing concentration affect the rate of a reaction?

For any chemical reaction involving solutions, the rate of reaction will depend on the concentration of at least one of the solutions. In any chemical reaction the concentration of reactants is the highest at the beginning and decreases as that reactant is used up as the reaction occurs.

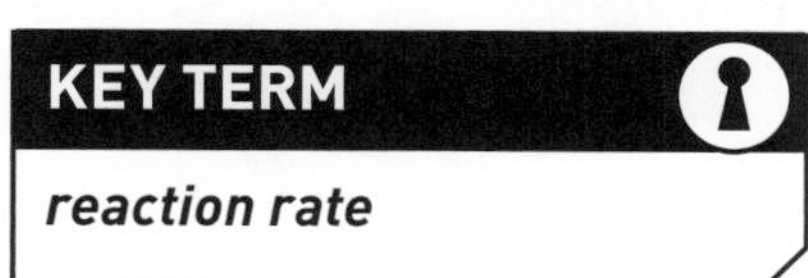

In this practical we will investigate the rate of reaction of magnesium metal reacting with dilute hydrochloric acid.

magnesium + hydrochloric acid → magnesium chloride + hydrogen

$$Mg(s) + 2HCl(aq) \rightarrow MgCl_2(aq) + H_2(g)$$

Aim

To investigate the effect of concentration on the rate of reaction between magnesium and hydrochloric acid by recording the time for the reaction to be completed at each concentration.

Apparatus and chemicals

- Eye protection
- 100 cm^3 beaker
- 50 cm^3 measuring cylinder
- Stopwatch
- Emery paper
- Glass rod
- Scissors
- Ruler
- Magnesium ribbon
- 2 mol/dm^3 hydrochloric acid solution

SAFETY GUIDANCE 

- Eye protection must be worn.
- The higher concentrations of hydrochloric acid are moderately harmful.

Method

Throughout the practical the student should wear eye protection.

1. Use the scissors to cut six pieces of magnesium ribbon exactly 2 cm long.
2. Use the emery paper to remove the magnesium oxide coating that forms on the pieces of magnesium ribbon.
3. Use a measuring cylinder to pour 40 cm^3 of 2 mol/dm^3 hydrochloric acid into the beaker.
4. Drop one of the pieces of magnesium ribbon into the acid and start the stopwatch. Use the glass rod to push the ribbon below the surface of the acid.
5. Stop the watch when the magnesium ribbon has completely reacted and can no longer be seen. Record the time, in seconds, in the table in the Observations section.
6. Rinse out the beaker and repeat for each of the other five acid concentrations made as shown in the table below. So for example, to make the 1.75 mol/dm^3 hydrochloric acid, pour 35 cm^3 of 2 mol/dm^3 hydrochloric acid into the measuring cylinder and add 5 cm^3 of water to make the total volume up to 40 cm^3.
7. Repeat for the remaining four hydrochloric acid concentrations.

Observations

Concentration of acid / mol/dm^3	Volume of 2 mol/dm^3 hydrochloric acid / cm^3	Volume of water / cm^3	Total volume / cm^3	Time / s	Rate / s
2	40	0	40		
1.75	35	5	40		
1.5	30	10	40		
1.25	25	15	40		
1	20	20	40		
0.5	10	30	40		

1 The rate of reaction can be found for this experiment using:

$$\text{Rate of reaction (/s)} = \frac{1}{\text{time (s)}}$$

Complete the table by finding the rate of reaction for the six different experiments.

2 Plot a graph showing the rate (/s) (y-axis) against the concentration of hydrochloric acid (mol/dm^3) (x-axis). Draw a smooth curve of best fit.

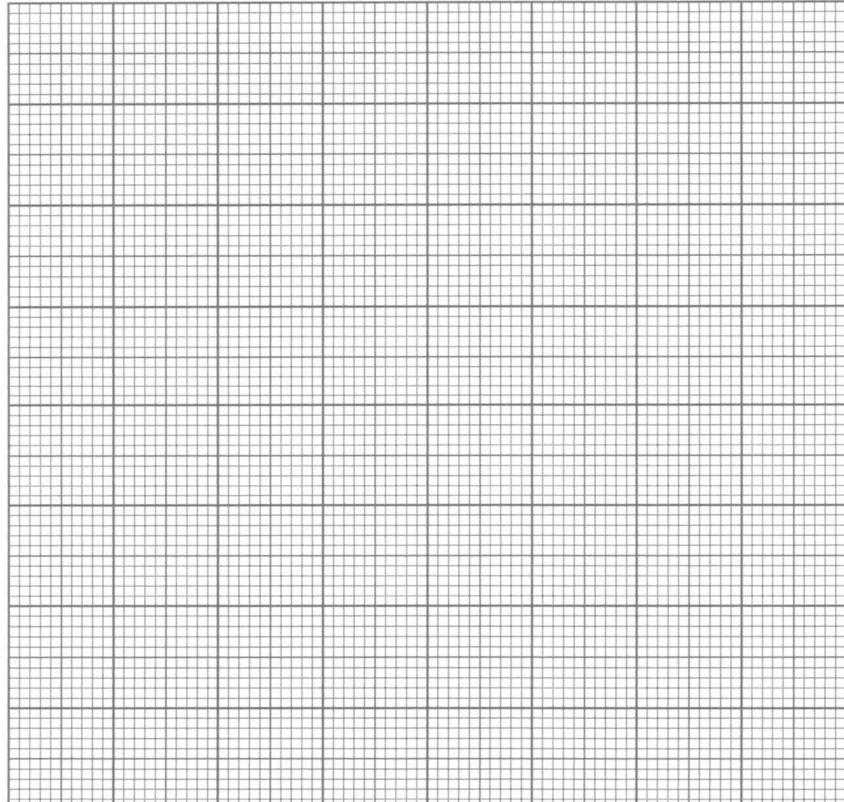

Conclusions

1 Which acid concentration causes the magnesium to react the fastest?

How can you tell this from the measurements you made?

....

....

2 In this reaction which of the reactants is used in excess?

How do you know this?

....

....

3 Use your graph to estimate the time taken for 2 cm of magnesium to react with 40 cm^3 of:

a 0.75 mol/dm^3 hydrochloric acid.

....

b 0.25 mol/dm^3 hydrochloric acid.

....

4 Describe how you could show that the gas produced was hydrogen.

....

....

5 Explain the findings in terms of particles and their collisions.

....

....

....

Evaluation

1 State the variables that you kept the same for each experiment.

....

....

2 What other factor, or factors, should be kept constant to show the effect of concentration on the rate of reaction? Were you able to control these?

....

....

3 Write down at least two ways in which this experiment could have been improved and state why they would have improved the experiment.

....

....

....

....

4 Describe one other way the rate of reaction could have been followed, other than by recording the time taken for the magnesium to fully react. Include the names of any apparatus required.

....

....

....

GOING FURTHER

Looking at your graph of rate against concentration of hydrochloric acid do you notice anything about the shape of the graph? The shape of the rate–concentration graph, from any experiment, can be used to assign an 'order of reaction'.

Use your research skills to find out about orders of reaction and how they can be found.

....

....

....

....

....

7.3 How does temperature affect the rate of a reaction?

The rate of any chemical reaction varies with temperature. To show the effect of a change in temperature we will use the reaction between hydrochloric acid and calcium carbonate, in the form of small marble chips.

KEY TERM

reaction rate

calcium carbonate + hydrochloric acid → calcium chloride + water + carbon dioxide

$$CaCO_3(s) + 2HCl(aq) \rightarrow CaCl_2(aq) + H_2O(l) + CO_2(g)$$

In this experiment the concentration of the hydrochloric acid remains constant but its temperature will be changed. The amount of calcium carbonate used will also remain the same. Any changes in the rate of the reaction will be due to the temperature at which the reaction is being carried out.

Aim

To investigate the effect of temperature on the rate of reaction between calcium carbonate and hydrochloric acid by recording the time needed to collect 50 cm^3 of carbon dioxide gas at each temperature.

Equipment

- Eye protection
- 100 cm^3 conical flask fitted with a bung and delivery tube
- 1 mol/dm^3 hydrochloric acid solution
- Six 1 g batches of small marble chips
- Small pieces of paper for adding marble chips to acid
- 50 cm^3 measuring cylinder
- 50 cm^3 burette
- Stopwatch
- Clamp stand and clamp
- Basin of water
- Thermometer
- Gauze
- Tripod
- Bunsen burner
- Top pan balance (minimum resolution 0.1 g)

TIP 

If a 50 cm^3 burette cannot be used, consider using a 50 cm^3 measuring cylinder under water. This will not be as accurate as using an inverted burette.

TIP 

It is important to be able to take a burette reading accurately. You must look at the bottom of the meniscus (the lowest part of the curve the liquid forms in the burette), at eye level to rule out parallax errors. In this experiment the scales will be upside down so take care when you are taking readings.

SAFETY GUIDANCE 

- Eye protection must be worn.
- Be careful when handling the heated conical flask. You should stand while carrying out the experiment and making measurements so that you can move away instantly if the flask containing the hot acid breaks or is knocked over.
- 1 mol/dm^3 hydrochloric acid solution is an irritant.

Method

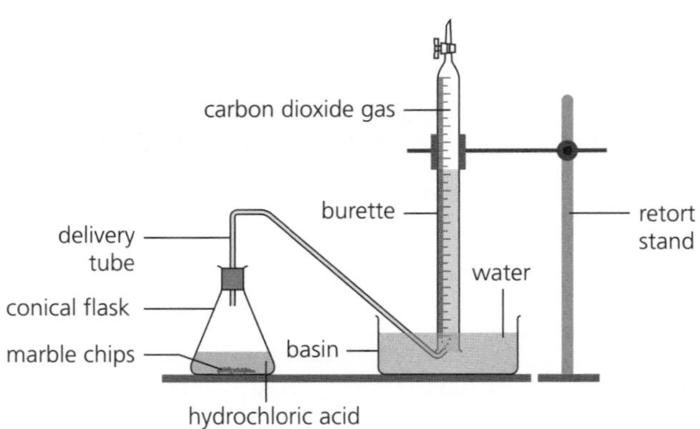

Figure 2

1 Use the top pan balance to weigh out, onto paper, six batches of small marble chips all with the same mass of 1 g.

2 Set up the burette as shown in Figure 2. You will need to fill the burette with water and then, with your finger over the open end, invert it in the basin of water. You may need to open the tap on the burette to lower the water level so that it starts on the scale on the burette.

3 Take the initial burette reading. You will be recording the time to collect 25.00 cm^3 of gas, so work out where the level of liquid will be when you need to stop the stopwatch.

4 Using the measuring cylinder pour 40 cm^3 of the 1 mol/dm^3 hydrochloric acid into the conical flask.

5 Place the conical flask onto a tripod and gauze and heat the acid to around 40 °C. When it has reached a temperature around 40 °C take an accurate measurement and record this as the actual starting temperature in the table in the Observations section.

6 To start the reaction, quickly add 1 g of marble chips to the flask and put the bung in. Start your stopwatch as soon as you add the marble chips to the flask, swirl the flask once and record in the table in the Observations section the time for 25.00 cm^3 of carbon dioxide gas to be collected.

7 Repeat for the other temperatures around 25 °C, 30 °C and 35 °C. In addition, carry out the experiment at room temperature.

TIP 

Fold each piece of paper so that it has a V-shape. This will help you to quickly add the marble chips to the acid at the start of the experiment.

TIP 

Burettes can be read to two decimal places, that is to the nearest 0.05 cm^3, where the second decimal place is either a 0, if the bottom of the meniscus sits on the scale division, or a 5 if it is between the divisions. So if, for example, the meniscus was between 49.10 cm^3 and 49.20 cm^3, the reading would be 49.15 cm^3.

The measuring cylinder would only be able to be read to the nearest 0.5 cm^3.

TIP 

It is better to work in pairs for this experiment as the next section needs to be done quickly but safely.

Observations

Expt. no.	Starting temperature / °C	Time / s	Rate of reaction / s
1			
2			
3			
4			
5			

1 The rate of reaction can be found for this experiment by using:

$$\text{Rate of reaction (/s)} = \frac{1}{\text{time (s)}}$$

Complete the table by finding the rate of reaction for the different experiments.

2 Plot a graph showing the rate (*y*-axis) against the temperature (*x*-axis). Draw a smooth curve between the points.

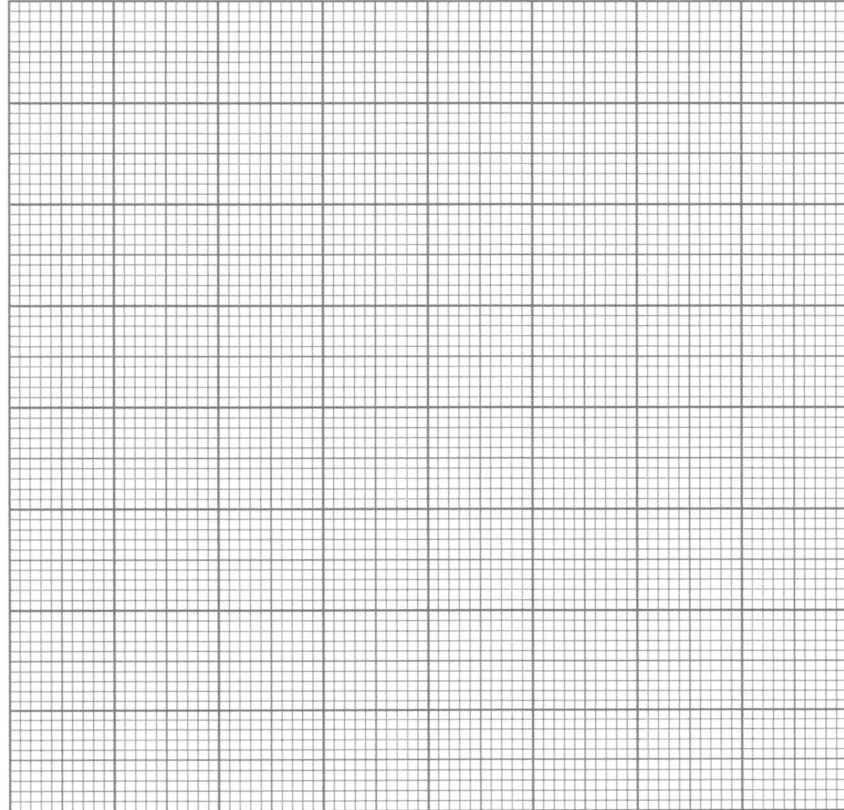

Conclusions

1 Describe the relationship between time for the reaction to be completed and the rate of the reaction.

....

....

2 What does your graph tell you about how temperature affects the rate of a reaction?

....

3 Explain your findings in terms of particles and collisions.

....

....

....

Evaluation

1 Identify two possible sources of error in the measurements in this experiment.

....

....

....

2 State the variables that you kept the same for the experiments.

....

....

....

3 What other factor, or factors, should be kept constant to show the effect of concentration on the rate of reaction? Were you able to control these?

....

....

4 Write down at least two ways in which this experiment could have been improved and state why they would have improved the experiment.

....

....

....

GOING FURTHER

At any given temperature the speeds, and therefore the kinetic energies, of the individual particles are widely spread. Use your research skills to find out about the Maxwell–Boltzmann distribution curve and suggest why the energies of the particles can be so very different.

....

....

....

8 Acids, bases and salts

8.1 Determination of the concentration of a solution of hydrochloric acid

The titration technique is used to find what volume of an acid *just* neutralises an alkali to produce a neutral solution. In this experiment we will use a solution of hydrochloric acid of unknown concentration and react it with a known volume of a solution of sodium hydroxide whose concentration is also known.

KEY TERMS

titration
neutralisation

hydrochloric acid + sodium hydroxide → sodium chloride + water

$$HCl(aq) \quad + \quad NaOH(aq) \quad \rightarrow \quad NaCl(aq) \quad + H_2O(l)$$

The end-point (when the acid just neutralises the alkali) is shown by an indicator.

From the balanced chemical equation, it can be seen that one mole of hydrochloric acid reacts with one mole of sodium hydroxide. Because we know the volume and concentration of the sodium hydroxide used and the volume of acid needed to neutralise the alkali, we will be able to work out the accurate concentration of the acid.

Aim

To find the end-point of the neutralisation reaction using the titration method.

Use the end-point to find the accurate concentration of a solution of hydrochloric acid.

Apparatus and chemicals

- Eye protection
- Burette
- Burette stand
- 25 cm^3 pipette and filler
- Filter funnel
- 250 cm^3 conical flask
- White tile
- Hydrochloric acid solution with a concentration of between 0.075 and 0.125 mol/dm^3
- 0.1 mol/dm^3 sodium hydroxide solution
- Thymolphthalein indicator

SAFETY GUIDANCE

- Eye protection must be worn.
- When you are putting the pipette filler onto the pipette, make sure that you hold the pipette close to the end the filler will be attached to. This should stop the pipette breaking as you push on the filler.
- Sodium hydroxide (0.1 mol/dm^3) – corrosive.
- Thymolphthalein – low hazard.
- Hydrochloric acid solution – irritant.

Method

Throughout the practical the student should wear eye protection.

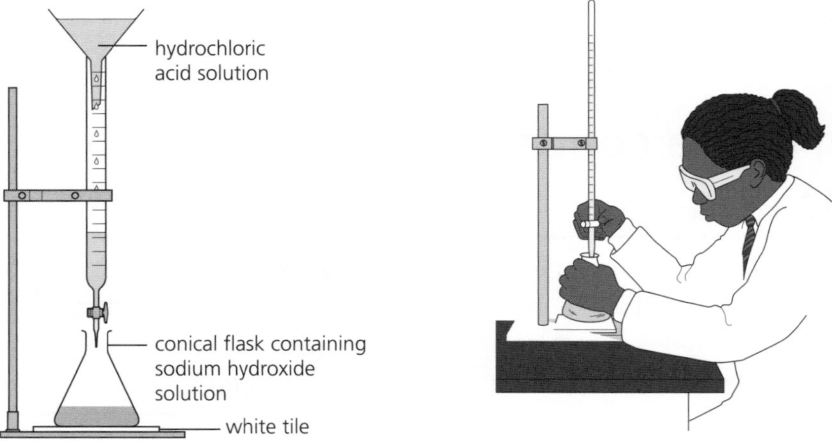

Figure 1 **Figure 2**

TIP

It is important to be able to take a burette reading accurately. You must look at the bottom of the meniscus (the lowest part of the curve the liquid forms in the burette), at eye level to rule out parallax errors. Burettes can be read to two decimal places, that is to the nearest 0.05 cm^3, where the second decimal place is either a 0, if the bottom of the meniscus is on the scale division, or a 5 if it is between the divisions. So if, for example, the meniscus was between 24.10 cm^3 and 24.20 cm^3 the reading would be 24.15 cm^3.

1 Clamp a burette vertically and use a filter funnel to fill it with some of the hydrochloric acid solution, ensuring that some of the acid solution runs through the tap (Figure 1). The filter funnel is now removed.

Record the initial reading of the burette to two decimal places, in the 'Rough' column in the table in the Observations section.

2 Using a pipette and filler, put 25.0 cm^3 of the 1 mol/dm^3 sodium hydroxide solution into a conical flask, on a white tile.

3 Add four drops of the indicator thymolphthalein to the sodium hydroxide solution in the conical flask. Make sure that you use the same number of drops each time.

4 Add small quantities of the acid from the burette – usually no more than 0.5 cm^3 at a time (Figure 2), swirling the flask as you do, until the indicator *just* changes colour from blue to colourless.

5 Record the final volume reading from the burette into the table in the Observations section to two decimal places, in the 'Rough' column. Work out the volume of acid you have used. This will give you an idea of where the end-point is and should allow you to add smaller quantities of acid around this point the next time you do it.

6 Repeat steps 2–4 until you have obtained three readings within 0.10 cm^3 of one another. Make sure you rinse out the conical flask with tap water between each experiment.

Observations

1 Complete the results table as you carry out your experiment.

Table 1

	Rough	1	2	3	4
Final burette reading / cm^3					
Initial burette reading / cm^3					
Volume of acid used / cm^3					

TIP

When you work out the average only use the three results within 0.10 cm^3 of each other. Do not use the 'Rough' value.

2 Work out an average volume (average titre) of acid used from the three results that were within 0.10 cm^3 of one another.

Average volume (titre) = ... cm^3

3 Calculate the number of moles of sodium hydroxide used in each experiment.

....

....

4 Find the number of moles of hydrochloric acid that would react with this number of moles of sodium hydroxide.

....

....

5 Using the average volume of hydrochloric acid used and the number of moles of it used in the reaction, calculate the concentration of the acid solution to two decimal places.

....

....

Conclusion

Does your answer fit in with the information given to you about the strength of the acid used?

....

....

Evaluation

1 Why is it important that the burette is clamped vertically?

....

2 What is the purpose of an indicator during a titration?

....

3 What is the purpose of the white tile?

....

4 Why should the same number of drops of thymolphthalein be used each time?

....

....

....

5 Why is it important to add the acid slowly to the alkali?

....

....

6 State some possible errors with the method used.

....

....

....

7 How could the procedure have been improved to give more accurate data?

....

....

GOING FURTHER

Use your research skills to find out and state how titrations are used in research or in the food industry.

....

....

8.2 Determination of the concentration of a solution of sodium hydroxide

KEY TERMS 

neutralisation

titration

Sulfuric acid is a cheap and excellent bulk acid. It is used extensively to make other substances such as detergents and fertilisers. It is also used by analytical chemists to determine unknown concentrations of alkalis. A titration is an important, and most widely used, analytical technique that is used in this situation. It is also used in forensic laboratories and water treatment laboratories, for example.

In titrations, the acid is used to *just* neutralise a known quantity of alkali.

In the case of sulfuric acid with sodium hydroxide:

hydrogen ions + hydroxide ions $\rightarrow$ water

$$2H^+(aq) \quad + \quad 2OH^-(aq) \quad \rightarrow 2H_2O(l)$$

If the volumes of the substances involved are known, as well as the concentration of the acid, then it is possible to calculate the unknown alkali concentration.

Aim

To use sulfuric acid in a titration to determine the concentration of a solution of sodium hydroxide.

Apparatus and chemicals

- Eye protection
- Burette
- Burette stand
- $2 \times 25\,cm^3$ pipettes and filler
- $250\,cm^3$ conical flask
- White tile
- Filter funnel
- Glass rod
- $0.05\,mol/dm^3$ sulfuric acid
- Sodium hydroxide solution (0.075–$0.125\,mol/dm^3$)
- Thymolphthalein indicator

SAFETY GUIDANCE

- Eye protection must be worn.
- When you are putting the pipette filler onto the pipette, make sure that you hold the pipette close to the end the filler will be attached to. This should stop the pipette from breaking as you push on the filler.
- Thymolphthalein indicator – low hazard
- Sodium hydroxide solution – corrosive
- Sulfuric acid solution – low hazard at this concentration

Method

Throughout the practical the student should wear eye protection.

1 Clamp a burette vertically and use a filter funnel to fill it with some of the sulfuric acid solution, ensuring that some of the acid solution runs through the tap (Figure 3). The filter funnel is now removed.

Record the initial burette reading to two decimal places in the 'Rough' column in the table in the Observations section.

2 Fill the pipette with $25.0\,cm^3$ sodium hydroxide. Pour this exact amount into a conical flask on a white tile. Add a few drops of thymolphthalein indicator to the sodium hydroxide solution in the conical flask. Thymolphthalein is blue in alkaline conditions but colourless in acid.

3 Add small quantities of the acid from the burette – usually no more than $0.5\,cm^3$ at a time (Figure 4), swirling the flask as you do, until the indicator *just* changes colour from blue to colourless.

4 Record the final volume reading from the burette into the table in the Observations section to two decimal places, in the 'Rough' column. Work out the volume of acid you have used. This will give you an idea of where the end-point is and should allow you to add smaller quantities of acid around this point the next time you do the titration.

TIP

It is important to be able to take a burette reading accurately. You must look at the bottom of the meniscus (the lowest part of the curve the liquid forms in the burette), at eye level to rule out parallax errors. Burettes can be read to two decimal places, that is to the nearest $0.05\,cm^3$, where the second decimal place is either a 0, if the bottom of the meniscus is on the scale division, or a 5 if it is between the divisions. So if, for example, the meniscus was between $24.10\,cm^3$ and $24.20\,cm^3$ the reading would be $24.15\,cm^3$.

5 Take the final reading on the burette at the end-point (just as neutralisation has taken place). Record this value in the table in the Observations section as the rough value.

6 Repeat steps 2–4 until you have obtained three readings within $0.10\,cm^3$ of one another. Make sure you rinse out the conical flask with tap water.

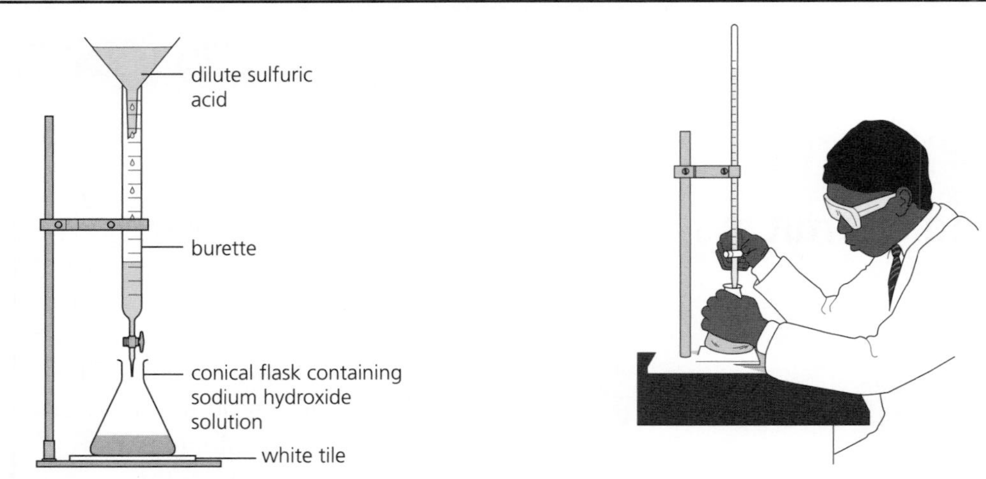

Figure 3

Figure 4

Observations

	Rough	1	2	3	4
Final burette reading / cm^3					
Initial burette reading / cm^3					
Volume of acid used / cm^3					

1 The average titre = cm^3

The chemical equation is:

sulfuric acid + sodium hydroxide → sodium sulfate + water

$H_2SO_4(aq) + 2NaOH(aq) \rightarrow Na_2SO_4(aq) + H_2O(l)$

2 Calculate the number of moles of sulfuric acid used.

....

3 From the balanced chemical equation you can see that one mole of sulfuric acid reacts with two moles of sodium hydroxide. You have already calculated the number of moles of sulfuric acid used, so how many moles of sodium hydroxide were present in the conical flask?

....

4 This number of moles of sodium hydroxide was present in 25.0 cm^3 of solution. How many moles would have been in 1000 cm^3 of solution?

....

Conclusion

From your results, state the concentration of the sodium hydroxide solution in mol/dm^3.

....

Evaluation

1 Why must the filter funnel be removed from the burette before titration can be carried out?

....

....

2 What is the purpose of the white tile?

....

3 Why do you have to swirl the contents of the conical flask as you add the acid?

....

4 Why is the first titration figure known as the 'rough' value?

....

....

5 How could the procedure have been improved to give more accurate data?

....

....

8.3 Preparation of hydrated magnesium sulfate

Magnesium sulfate can be obtained from the reaction between magnesium carbonate and sulfuric acid.

Once the magnesium sulfate solution is formed, the hydrated salt is obtained as a solid by evaporation of some of the water, allowing some water to remain and be taken into the crystals. The hydrated crystals, once formed, can then be dried further.

The water incorporated into the crystal structure of a hydrated salt is referred to as water of crystallisation

KEY TERMS 

salt
hydrated salt
crystallisation
filtration
filtrate
soluble
insoluble

water of crystallisation

magnesium carbonate + sulfuric acid → magnesium sulfate + water + carbon dioxide

$$MgCO_3(s) + H_2SO_4(aq) \rightarrow MgSO_4(aq) + H_2O(l) + CO_2(g)$$

Aim

To prepare hydrated magnesium sulfate crystals from the reaction between sulfuric acid (an acid) and magnesium carbonate (a carbonate).

Apparatus and chemicals

- Eye protection
- $150\,cm^3$ beaker
- $250\,cm^3$ beaker
- $25\,cm^3$ measuring cylinder
- Glass rod
- Spatula
- Filter funnel
- 2 × filter paper
- Evaporating basin
- Tripod
- Gauze
- Bunsen burner
- Heat-proof mat
- Magnesium carbonate solid
- $1\,mol/dm^3$ sulfuric acid solution

SAFETY GUIDANCE 

- Eye protection must be worn.
- Allow the hot evaporating basin to cool before removing it from the tripod.
- Do not allow the magnesium sulfate solution to boil dry during crystallisation.
- Magnesium carbonate solid – irritant
- $1\,mol/dm^3$ sulfuric acid solution – irritant
- Hydrated magnesium sulfate – low hazard

Method

Throughout the practical the student should wear eye protection.

1. Using a measuring cylinder, measure out $25\,cm^3$ of sulfuric acid into the $150\,cm^3$ beaker.
2. Add the magnesium carbonate crystals to the acid, using the spatula, stirring gently with a glass rod.
3. When the effervescence has stopped and you can see some magnesium carbonate crystals at the bottom of the beaker, stop adding the magnesium carbonate.
4. Filter the contents of the beaker, using a funnel and filter paper, allowing the filtrate to run into an evaporating basin until it is half full (Figure 5). The residue will collect in the filter paper.

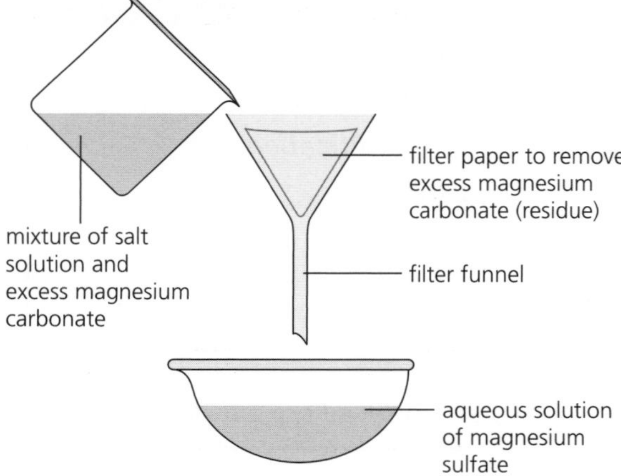

Figure 5

5 Place the evaporating basin onto a 250 cm^3 beaker, as shown in Figure 6. Light the Bunsen burner and evaporate the filtrate slowly until there is only a small amount of solution left. Do not allow the solution to boil dry.

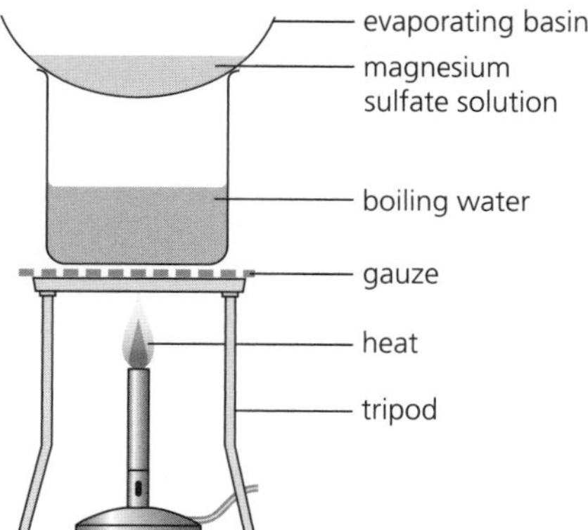

Figure 6

6 Place another piece of filter paper over the evaporating basin.

7 Leave the remainder of the solution to cool and allow crystallisation to occur over the next few days.

Observation

What colour are the hydrated magnesium sulfate crystals you have produced?

....

Conclusions

1 Is hydrated magnesium sulfate a soluble or insoluble salt?

....

2 Why is water needed to allow the crystals to form?

....

....

3 While heating the solution in the evaporating basin you may have noticed some powdery crystals forming around the edges of the basin. What do you think this powder was and how do you think it was formed?

....

....

Evaluation

1 What caused the effervescence?

....

2 Why did you stop adding the magnesium carbonate when the effervescence stopped?

....

....

3 Why is it important to be able to see some magnesium carbonate at the bottom of the beaker in step 3?

....

....

4 Why was only half of the volume of the solution evaporated?

....

....

5 What was the purpose of the filter paper being placed over the evaporating basin in step 6?

....

....

6 A student was told to prepare a sample of calcium chloride from solid calcium carbonate.

Describe how the student should obtain crystals of calcium chloride. Include what acid to use, and the process for obtaining pure, dry crystals of hydrated calcium chloride.

....

....

....

....

....

8.4 Properties of dilute sulfuric acid

Dilute sulfuric acid has the typical properties of an acid in that it will react with and be neutralised by alkalis, metal oxides, metals and carbonates to produce salts known as sulfates.

Dilute sulfuric acid is a typical strong acid. Sulfuric acid is made by diluting concentrated sulfuric acid with water. When water is added to the acid, complete dissociation takes place as the following process occurs:

sulfuric acid $\xrightarrow{\text{water}}$ hydrogen ions + sulfate ions

$$H_2SO_4(aq) \xrightarrow{\text{water}} 2H^+(aq) + SO_4^{2-}(aq)$$

KEY TERMS 

indicator
pH scale
neutralisation
strong acid
weak acid
ionisation
complete dissociation

Aim

To look at the effect of dilute sulfuric acid on some common indicators.
To examine the properties and reactions of dilute sulfuric acid.

To look at the differences between a strong and weak acid in their reaction with magnesium metal.

Apparatus and chemicals

- Eye protection
- Test-tube rack
- Test tubes
- Thermometer (–10 to 110 °C)
- Delivery tube
- 1 mol/dm^3 sulfuric acid
- 1 mol/dm^3 ethanoic acid
- Wooden spills
- 3 × 1 cm pieces magnesium ribbon
- Copper(II) oxide powder
- Spatula
- Sodium carbonate solution
- 0.4 mol/dm^3 sodium hydroxide solution
- Limewater
- Glass rods
- Short lengths of red litmus paper, blue litmus paper and universal indicator paper
- Universal indicator colour chart
- Methyl orange indicator solution
- Thymolphthalein indicator solution
- Dropping pipettes
- Cavity tile (also known as a spotting tile or dimple tile)

SAFETY GUIDANCE 

- Eye protection must be worn.
- Sulfuric acid (1 mol/dm^3) – irritant
- Ethanoic acid (1 mol/dm^3) – irritant
- Magnesium ribbon – low hazard
- Copper(II) oxide – harmful, dangerous for the environment
- Sodium hydroxide solution (0.4 mol/dm^3) – irritant
- Sodium carbonate – low hazard
- Limewater – irritant
- Thymolphthalein – low hazard
- Methyl orange – toxic, corrosive, irritant, dangerous for the environment

Method 1

Throughout the practical the student should wear eye protection.

In the first part of the experiment you are going to look at the colour changes that occur when a solution of dilute sulfuric acid and a solution of dilute sodium hydroxide solution are added to some common indicators.

1 Place four drops of methyl orange indicator and four drops of thymolphthalein indicator in separate holes on the cavity tile.

2 Tear small pieces off the litmus paper and the universal indicator paper and place in separate holes on the cavity tile.

3 Using a dropping pipette, add a few drops of sulfuric acid to each hole containing an indicator. Record your observations in the table in the Observations section.

4 Repeat these tests using the dilute sodium hydroxide solution, making sure that you have rinsed off the cavity tile and dropping pipettes with tap water.

Observations

Indicator	Colour of the indicator with dilute sulfuric acid	Colour of the indicator with dilute sodium hydroxide
Red litmus paper		
Blue litmus paper		
Universal indicator paper		
Methyl orange		
Thymolphthalein		

Method 2

1 Place 1 cm magnesium ribbon into a test tube and add a 2 cm depth of sulfuric acid to it. Put your thumb over the end to stop any gas escaping and test the collected gas with a lighted spill.

Write your observations in the table in the Observations section.

2 Place a small sample of the sodium hydroxide solution into a test tube in the test-tube rack. Insert the thermometer and record the temperature. Add a small amount of sulfuric acid to it. Measure and record the temperature as the reaction occurs.

Write your observations in the table in the Observations section.

3 Place a little sodium carbonate into a test tube and add a small amount of sulfuric acid to it. Put the delivery tube into the neck and pass any gas through limewater.

Write your observations in the table in the Observations section.

4 Add some hot water from a kettle to a beaker. Place a test tube containing 3 cm depth of dilute sulfuric acid into the beaker of hot water to warm up the acid. Place 1 spatula of copper(II) oxide into a test tube.

Write your observations in the table in the Observations section.

Observations

Complete each row of the table by writing in the name of the salt produced in each case.

Substance	Reaction with dilute sulfuric acid	Explanation of your observations	Name of salt produced
Magnesium			
Sodium hydroxide			
Sodium carbonate			
Copper(II) oxide			

Method 3

In the final part of the experiment you will look at the difference between sulfuric acid and ethanoic acid (CH_3COOH).

1 Half fill two test tubes to the same height: one with 1 mol/dm^3 sulfuric acid solution and one with 1 mol/dm^3 ethanoic acid solution.

2 Put a 1 cm length of magnesium ribbon into each tube at the same time and use a glass rod in each tube to ensure that they are kept from floating on the surface of the acid. Make a note of your observations in the table in the Observations section.

3 Finally tear off two small pieces of universal indicator paper and place them in two separate holes of the cavity tile.

4 To one of the holes add two drops of the 1 mol/dm^3 sulfuric acid solution and to the other hole add two drops of 1 mol/dm^3 ethanoic acid solution. Record your observations in the table in the Observations section.

Observations

Acid	Observations	Tick if fastest to react with the magnesium (✓)	Colour of universal indicator paper
Sulfuric acid			
Ethanoic acid			

Conclusions

1 Write a word equation for the reaction of sulfuric acid with:

a magnesium

....

b sodium hydroxide

....

c sodium carbonate

....

d copper(II) oxide

....

2 Deduce the symbol equation, with state symbols, for the reaction of sulfuric acid with:

a magnesium

....

b sodium hydroxide, NaOH(aq)

....

c sodium carbonate, Na_2CO_3(aq)

....

d copper(II) oxide

....

3 What would be formed if the gas produced in step 2 of Method 3 is tested with a lighted spill and it reacts with the oxygen in the air?

....

4 What is the evidence that dilute sulfuric acid has the typical properties of an acid?

....

....

5 Write a word equation for the reaction of ethanoic acid with magnesium.

....

6 Deduce the symbol equation, with state symbols, for the reaction of ethanoic acid with magnesium.

....

7 What is the approximate pH of:

a 1 mol/dm^3 sulfuric acid solution?

....

b 1 mol/dm^3 ethanoic acid solution?

....

8 Explain why one of the acids was faster than the other in its reaction with magnesium.

....

....

....

....

....

....

Evaluation

1 Why do you put your thumb over the test tube in the reaction between magnesium and sulfuric acid?

....

2 How could you tell in step 4 of Method 2 that a chemical reaction had taken place?

....

3 Why is a delivery tube needed in the reaction between sodium carbonate and sulfuric acid?

....

4 How can you be sure that the comparison of sulfuric acid and ethanoic acid in Method 3 was only due to them being different acids?

....

....

5 Outline how this experiment could be improved to compare the relative acidity of the two acids more accurately.

....

....

GOING FURTHER

It is often said that the countries which produce the most sulfuric acid are the richest countries in the world. Use your research skills to find and state the top five sulfuric acid producing countries. Do your findings show that the statement is true? Why is sulfuric acid such an important chemical?

....

....

....

....

....

9 The Periodic Table

9.1 Reactions of the Group 1 metals (Teacher demonstration)

The metals in Group 1 of the Periodic Table are known as the alkali metals. This is because when they react with water they all form the metal hydroxide, producing alkaline solutions. For example, when lithium reacts with water, lithium hydroxide is formed:

lithium + water → lithium hydroxide + hydrogen

$2Li(s) + 2H_2O(l) \rightarrow 2LiOH(aq) + H_2(g)$

The two alkali metals available for use in school are lithium and sodium. They are both very reactive metals and are stored under oil to prevent them from reacting with water and oxygen in the air.

All of the Group 1 metals are soft metals and are easily cut with a knife. When they are cut, they show a bright silvery colour which rapidly tarnishes on exposure to the air.

KEY TERMS 

alkali metal
group
electronic configuration

SAFETY GUIDANCE 

- Eye protection must be worn.
- Sodium and lithium are highly flammable and corrosive. Both of these metals are very reactive with water, hence place a safety screen close to the trough. Students should wear eye protection and be 3 m away. Only small pieces of the metals should be used and advice from CLEAPSS should be followed as to the appropriate size and storage methods which should be used (Li: HC058a, Na: HC088).
- Solutions of the hydroxides of sodium and lithium – likely to be an irritant at the concentrations produced in the demonstration.

Aim

To demonstrate the reactivity of the Group 1 metals lithium and sodium by looking at their reaction with water. To be able to use the observations obtained to predict the reactions of the other Group 1 metals.

Apparatus and chemicals

- Eye protection
- Safety screen
- 2 × water troughs
- 2 × scalpels
- 2 × tweezers
- 2 × boiling tubes
- 2 × dropping pipettes
- White tile
- Universal indicator solution and chart
- Boiling-tube rack
- Filter paper
- Sodium metal in the original container
- Lithium metal in the original container

Method

Throughout the practical the teacher and all students should wear eye protection.

1 Half fill the two water troughs and place them side by side on the bench, behind a safety screen.

2 Using tweezers, take out a piece of lithium metal from its container and place it on a dry white tile. Hold the metal with some tweezers and use a scalpel to cut off a small piece, not more than 0.5 cm^3. Replace the remainder of the lithium metal back into its container, under oil.

3 Show the students the newly cut surface of the metal. They should record their observations in the table in the Observations section.

4 Use a piece of filter paper to remove some of the oil from the piece of metal you have cut off.

5 Use the tweezers to drop the lithium into the middle of the water in one of the water troughs. Students should record their observations in the table in the Observations section.

6 When the reaction is finished, remove some of the solution formed using a dropping pipette and put it into a boiling tube.

7 Add a few drops of universal indicator solution to the solution in the tube and ask students to record the colour and the pH of the solution formed in the table in the Observations section.

8 Repeat steps 1–7 for sodium, using a different water trough.

Observations

Complete this table as your teacher carries out the demonstration.

Metal	Colour of metal	Observations when added to water	Colour of solution formed with universal indicator solution	pH of the solution formed
Lithium				
Sodium				

Conclusions

1 From your observations, what can be said about the density of the two metals?

....

....

2 Which of the metals was, based on your observations, the most reactive?

....

3 Explain why this metal was the more reactive.

....

....

....

4 Write a balanced symbol equation for the reaction of sodium metal with water.

....

5 Why do both of the solutions produced have a pH above 7?

....

....

6 Why were the other four Group 1 metals, potassium, francium, caesium and rubidium, not used in the experiment?

....

....

7 Which of potassium, francium, caesium and rubidium would be the most reactive? Explain your prediction.

....

....

Evaluation

1 What is the purpose of the safety screen?

....

....

2 Why should the remainder of the metal which is *not* going to be used in the experiment be returned to its container?

....

....

3 What happened to the newly cut surfaces of the metals you were shown? Explain why this happens.

....

....

....

4 Why was only a small piece of each metal used?

....

....

5 Why was the filter paper used to remove most of the oil from the surface of each metal?

....

....

....

6 Why was each metal placed into the middle of the water in the trough?

....

....

....

....

7 In order to describe the difference in reactivity of the two metals used in the experiment you used your observations. Did the procedure used provide enough evidence of this?

....

....

8 Describe another way in which numeric data could be produced to show the trend in reactivity.

....

....

GOING FURTHER

Find a use for each of the first three Group 1 metals, or their compounds.

....

....

9.2 Halogen displacement reactions

When a more reactive metal reacts with a compound of a less reactive metal, a displacement reaction will occur, with the more reactive metal displacing the less reactive metal from the salt. The same idea works for the Group 7 elements. If a more reactive halogen reacts with a compound of a less reactive halogen, the less reactive halogen will be displaced and it will form the halogen molecule, while the more reactive halogen becomes a halide ion.

KEY TERMS 

displacement reaction

halogen

To decide whether a halogen displacement reaction has taken place you will need to look for a colour change occurring when the halogen in solution and halide ion are mixed together.

Aim

Displacement reactions can be used to determine the reactivity of the Group 7 elements, the halogens. In this experiment you will use these reactions to determine an order of reactivity for iodine, chlorine and bromine. You will then use this to predict the reactivity of the other two halogens, fluorine and astatine.

Apparatus and chemicals

- Eye protection
- 6 × test tubes
- 6 × test-tube bungs
- Dropping pipettes
- Test-tube rack
- 0.01 mol/dm^3 chlorine water
- 0.01 mol/dm^3 bromine water
- 0.01 mol/dm^3 iodine water
- 1 mol/dm^3 sodium chloride solution
- 1 mol/dm^3 sodium bromide solution
- 1 mol/dm^3 sodium iodide solution
- Beaker of sodium thiosulfate solution, placed in fume cupboard, for waste solutions.

SAFETY GUIDANCE 

- Eye protection must be worn.
- The contents of all your test tubes should be emptied into a beaker of sodium thiosulfate solution (low hazard) in a fume cupboard, not poured down the sink. If no fume cupboard is available, make sure the room is well ventilated.
- Avoid breathing in fumes. Take special care if you are asthmatic.
- Chlorine water – low hazard
- Bromine water – low hazard
- Iodine water – harmful
- 1 mol/dm^3 sodium chloride solution – low hazard
- 1 mol/dm^3 sodium bromide solution – low hazard
- 1 mol/dm^3 sodium iodide solution – low hazard

Method

Throughout the practical the student should wear eye protection.

1. Using a clean test tube and dropping pipette, place 1 cm depth of chlorine water into the tube.
2. Using a different pipette, add into the same tube a 1 cm depth of sodium iodide solution.
3. Place a bung in the tube and shake it.
4. If you see a change in colour write in the table in the Observations section that a 'Reaction has occurred'. If no colour change is seen write 'No reaction'.
5. Repeat steps 1–2 using the solution combinations shown below, using fresh test tubes each time.
   - **a** bromine water and sodium iodide
   - **b** chlorine water and sodium bromide
   - **c** iodine water and sodium bromide
   - **d** bromine water and sodium chloride
   - **e** iodine water and sodium chloride

Observations

Complete this results table as you carry out the experiment.

	Chlorine water	Bromine water	Iodine water
Colour after shaking with sodium iodide solution			
Colour after shaking with sodium bromide solution			
Colour after shaking with sodium chloride solution			

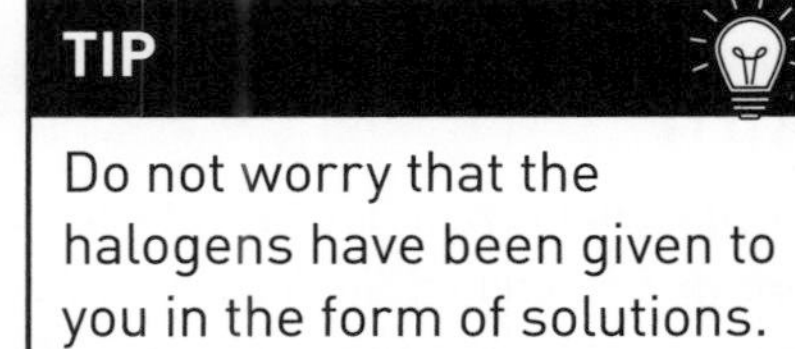

TIP

Do not worry that the halogens have been given to you in the form of solutions.

Conclusions

1 In how many of the six reactions you carried out has a displacement reaction occurred?

....

2 For each of these displacement reactions, write a balanced symbol equation.

....

....

....

3 What type of chemical change has happened to the halogen molecules in the displacement reactions?

....

4 Deduce the order of reactivity of the three halogens used in the experiment.

....

5 What would be the colour of fluorine and astatine in water?

....

....

Evaluation

1 Why is it important to use a bung on the tube, rather than putting your thumb over the open end before shaking?

....

....

2 Why were you given solutions of the halogens to use, rather than the halogens themselves?

....

....

....

3 Why was fluorine not used in this experiment?

....

4 Why was astatine not used in this experiment?

....

5 Did the experiment clearly show you the order of reactivity of the three halogens used?

....

....

GOING FURTHER

1 Use your research skills to find out a use for each of the five halogens or their compounds.

....

....

....

2 Which of all the five halogen elements is the most reactive? Explain your answer.

....

....

9.3 Using transition metal ions as catalysts

KEY TERMS 

reaction rate

catalyst

The iodine 'clock' reaction is a reaction between two colourless solutions, potassium iodide and potassium peroxodisulfate.

potassium iodide + potassium peroxodisulfate → potassium sulfate + iodine

$$2KI(aq) \quad + \quad K_2S_2O_8(aq) \quad \rightarrow \quad 2K_2SO_4(aq) \quad + I_2(aq)$$

When the reaction occurs, iodine is formed. In the presence of iodine indicator (or starch), the production of the iodine results in the formation of a dark blue/black colour.

The reaction is, however, so fast that another chemical, sodium thiosulfate, is added to slow down the formation of the blue/black colour. The sodium thiosulfate reacts with the iodine as it is produced by the reaction, so there is no free iodine to turn the indicator blue/black. When all the sodium thiosulfate has been used up the iodine reacts with the indicator producing the blue/black colour.

You will first carry out the reaction without a catalyst and then add the five potential catalyst solutions, one at a time, to the reaction mixture, to see whether they speed up the formation of the blue/black colour and increase the rate of the reaction.

The rate of the reaction can be found using:

$$\text{reaction rate (/s)} = \frac{1}{\text{time (s)}}$$

because the same amount of iodine is being produced each time by the reaction to bring about the formation of the blue/black colour.

Aim

To find transition metal compounds which will act as catalysts for the iodine clock reaction.

Apparatus and chemicals

- Eye protection
- 12 × boiling tubes
- 3 × 10 cm^3 measuring cylinders
- Boiling-tube rack
- Spatula
- Stopwatch
- 5 × dropping pipettes
- 0.5 mol/dm^3 potassium iodide solution
- 0.04 mol/dm^3 potassium peroxodisulfate solution
- 0.01 mol/dm^3 sodium thiosulfate solution
- Iodine indicator (or freshly made starch solution)
- 0.1 mol/dm^3 ammonium iron(III) sulfate solution
- 0.2 mol/dm^3 ammonium molybdate solution
- 0.2 mol/dm^3 chromium(III) chloride solution
- 0.2 mol/dm^3 iron(II) sulfate solution
- 0.2 mol/dm^3 copper(II) sulfate solution

SAFETY GUIDANCE

- Eye protection must be worn.
- Potassium peroxodisulfate solution – harmful, irritant.
- Sodium thiosulfate solution – low hazard.
- Iodine indicator (or freshly made starch solution) – low hazard.
- Ammonium iron(III) sulfate solution – low hazard.
- Ammonium molybdate solution (2 mol/dm^3) – irritant.
- Chromium iron(III) chloride solution (0.2 mol/dm^3) – low hazard.
- Iron(II) sulfate solution (0.2 mol/dm^3) – low hazard.
- Copper(II) sulfate (0.2 mol/dm^3) – harmful, irritant.

Method

Throughout the practical the student should wear eye protection.

1. Using a measuring cylinder, measure out 5 cm^3 of the potassium iodide solution and pour it into a boiling tube.
2. Into the same boiling tube, but using a different measuring cylinder, pour in 2 cm^3 of the sodium thiosulfate solution.
3. Into the same boiling tube, add a spatula full of iodine indicator powder (or 1 cm^3 of starch solution).

TIP

Remember to wash out your measuring cylinders with tap water so that you do not put the wrong chemical into the wrong cylinder.

4 Into a different boiling tube, using another measuring cylinder, place 2 cm^3 of potassium peroxodisulfate solution.

5 When you are ready to start the experiment, pour the tube containing the potassium peroxodisulfate solution into the other tube, swirl once, place it back into the rack and start the stopwatch.

6 Stop the stopwatch when you observe the formation of the blue/black colour. Record the time in the table in the Observations section, as accurately as your stopwatch allows. This is the time for the uncatalysed reaction.

7 You will now carry out exactly the same experiment, but this time you will add 10 drops of a potential catalyst solution to the tube containing the potassium peroxodisulfate, using a dropping pipette. Each time, use the table in the Observations section to record the time taken for the blue/black colour to form.

Observations

1 Complete the results table as you carry out the experiment.

Catalyst	Time / s	Rate / s
Uncatalysed reaction		
Iron(III) sulfate		
Ammonium molybdate		
Chromium(III) chloride		
Iron(II) sulfate		
Copper(II) sulfate		

2 Work out the rate for each of the six experiments and write it in the table.

3 Plot a bar chart to show the rate of reaction for each of the six reactions you have carried out.

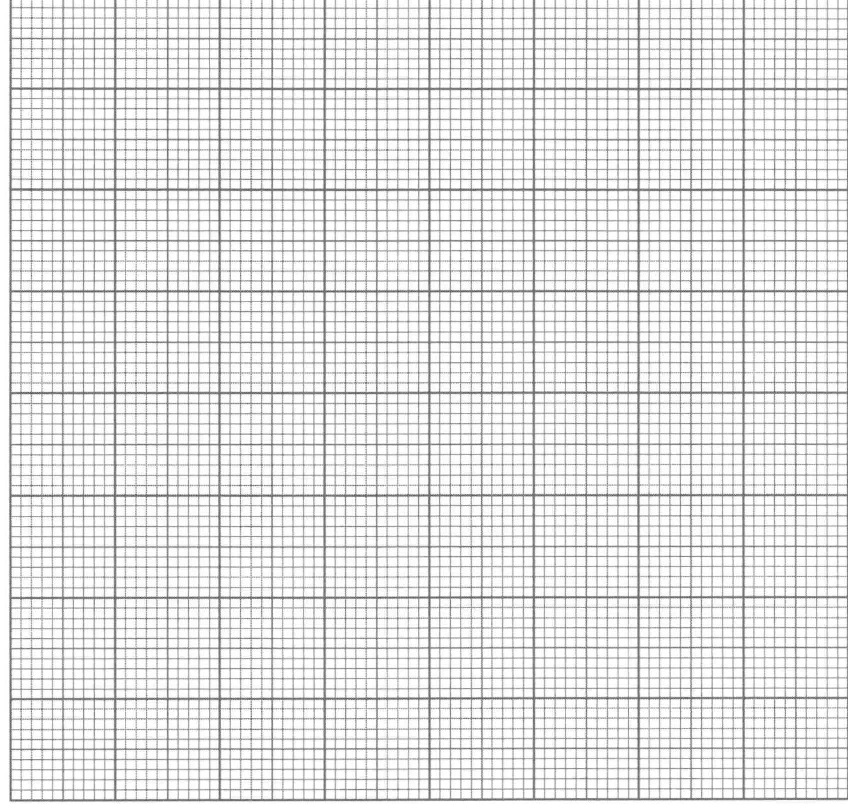

Conclusions

1 Which of the transition metal compounds acted as a catalyst for the reaction?

....

2 How do you know which of the compounds was the best catalyst?

....

3 Did any of the compounds slow the rate of the reaction down?

....

4 Apart from speeding up a chemical reaction, what is the other important factor when defining a compound as a catalyst?

....

5 Explain how a catalyst increases the rate of a chemical reaction.

....

....

....

....

....

Evaluation

1 Why is it important to use different measuring cylinders for each of the solutions?

....

....

TIP

Find the formula of each catalyst, from the charges on the ions, and compare the relative amounts of metal ion in each compound.

2 Why is the concentration of the iron(III) sulfate half that of the other catalysts?

....

....

....

....

3 Why is the potassium peroxodisulfate kept separate from the other chemicals in a different boiling tube?

....

....

4 Why is it important to swirl the tube once only for each experiment?

....

....

5 Suggest how the procedure could have been improved to get more accurate results. For each of your suggestions, state why they would improve the procedure.

....

....

....

....

GOING FURTHER

Use your research skills to find out how catalysts are tested in the laboratory by companies developing potential new catalysts.

....

....

....

....

....

....

10 Metals

10.1 Metal reactivity

The majority of the elements we know are metals. Metallic elements have characteristic physical properties. In other ways they show very different properties, particularly in their chemical properties. Iron, for example, will rust quickly if left unprotected, while gold remains totally unchanged after hundreds of years. Iron is said to be reactive in comparison to gold, which is unreactive.

KEY TERM

reactivity series of metals

In this experiment, you will be given four different metals and some hydrochloric acid solution. You will use the reaction between the metal and acid to produce an order of reactivity for the four metals. Metals react with hydrochloric acid solution to give the metal chloride and hydrogen gas:

metal + hydrochloric acid $\rightarrow$ metal chloride + hydrogen

You will need to ensure that you use the same mass of each metal in each experiment.

Aim

To use the reaction of different metals with acid to produce an order of reactivity for metals.

Apparatus and chemicals

- Eye protection
- 4 × boiling tubes
- Boiling-tube rack
- Top pan balance (minimum resolution 0.01 g)
- 25 cm^3 measuring cylinder
- Stopwatch
- Access to filings of magnesium, copper, iron and zinc
- Spatula
- 2 mol/dm^3 hydrochloric acid solution

SAFETY GUIDANCE 

- Eye protection must be worn.
- Hydrochloric acid solution – irritant
- Iron filings – low hazard
- Zinc (filings) – low hazard
- Do not hold the boiling tubes during the reactions because some of the reactions are highly exothermic and the tubes will get very hot.

Method

Throughout the practical the student should wear eye protection.

1. Using the measuring cylinder, put 10 cm^3 of the hydrochloric acid into a boiling tube and place it into the boiling-tube rack.
2. Starting with any one of the four metals, weigh out 0.2 g of the metal filings.
3. When you are ready to start the experiment, add the metal filings to the acid and start the stopwatch. Swirl the contents of the tube once.
4. Record the time it takes for all of the metal to react. If it takes over 5 minutes, record this in the table in the Observations section.
5. Repeat steps 1–4 for the other three metals, recording your results for each one.

Observations

Complete the table as you carry out the experiment.

Metal	Observation (s)	Time / s
Magnesium		
Copper		
Iron		
Zinc		

Conclusions

1 Which of the metals did not react at all with the hydrochloric acid?

....

2 Which of the metals was the most reactive? Give a reason for your answer.

....

3 Using your results, write down the order of reactivity you have obtained, starting with the least reactive metal.

....

4 Write a word and a balanced chemical equation for the reaction between zinc and hydrochloric acid.

....

....

5 In the reaction between zinc and hydrochloric acid, which component is oxidised and which is reduced?

....

....

6 Write two ionic half-equations for the reaction between zinc and hydrochloric acid.

....

....

7 Why is one metal more reactive than another in this type of reaction?

....

....

Evaluation

1 State the variables that you kept the same for each experiment.

....

....

....

2 Explain why it was important that the metals were all filings.

....

....

3 How could the method be improved to give a more quantitative comparison?

....

....

....

4 Use your research skills to find the names of three other metals which would *not* have reacted with the hydrochloric acid.

....

....

10.2 Metal displacement reactions

A more reactive metal will displace a less reactive metal from a solution of its salt. For example, copper is a more reactive metal than silver and so would displace silver from its salt:

copper + silver nitrate $\rightarrow$ copper(II) nitrate + silver

$Cu(s) + 2AgNO_3(aq) \rightarrow Cu(NO_3)_2(aq) + 2Ag(s)$

A reaction is known to have occurred because a colour change is observed. As the reaction proceeds, the solution turns blue as copper(II) nitrate is formed and the pink colour of copper metal is replaced by a more silvery solid product, silver. The change in colour is a way of indicating whether a reaction has occurred or not.

KEY TERMS 

displacement reaction
reactivity series of metals

redox reaction

Aim

To use displacement reactions to place four metals, magnesium, copper, iron and zinc, into an order of reactivity.

Apparatus and chemicals

- Eye protection
- Cavity tile with 16 holes, or two cavity tiles (or 16 test tubes)
- 4 × dropping pipettes
- Cleaned small pieces (0.5 cm × 0.5 cm) of magnesium, copper, iron and zinc
- Solutions (1 mol/dm^3) of the nitrates (or other suitable solutions) of magnesium, copper, iron and zinc

SAFETY GUIDANCE 

- Eye protection must be worn.
- Magnesium nitrate – harmful.
- Copper(II) nitrate – harmful.
- Iron(II) nitrate – harmful.
- Zinc nitrate – harmful.

Method

Throughout the practical the student should wear eye protection.

1 Using a dropping pipette, fill up four of the holes in the tile (or 1 cm depth in six test tubes) with magnesium nitrate solution.

2 Add a small piece of each metal to separate holes (or test tubes). Ensure that the metal is pushed under the solution.

3 Leave the reaction mixtures for 5 minutes.

4 Look at each of the reaction mixtures. Where there has been a change of colour on the surface of the metal or in the colour of the solution, you can assume that a reaction has occurred. In this case, place a tick (✓) in the table in the Observations section. Where no colour change has occurred, no reaction will have happened so place a cross (✗) in the table in the Observations section.

5 Repeat the process for the other three nitrate solutions, using a clean pipette each time.

Observations

Complete this table as you carry out your experiments.

	Metal nitrate solutions			
Metals	**Magnesium**	**Copper**	**Iron**	**Zinc**
Magnesium				
Copper				
Iron				
Zinc				

TIP 

Bubbles of gas (hydrogen) may also be observed as some of the metal nitrate solutions are acidic, but you should ignore any bubbles as they are not part of the displacement reactions.

Conclusions

1 Of the metals used, which reacted most often?

....

2 Of the metals used, which reacted least often?

....

3 Using your results, place the metals in a list with the most reactive metal first.

....

4 Write a balanced symbol equation for the reaction between magnesium and copper(II) sulfate.

....

5 In all the displacement reactions that have occurred, what chemical process is happening to each of the more reactive metals? State why you have reached this conclusion.

....

....

6 Write an ionic half-equation to show this change for copper in its reaction with silver nitrate solution.

....

7 What is happening to each of the less reactive metals in the displacement reactions? Explain your answer.

....

....

8 Write an ionic half-equation to show this change for the silver present in silver nitrate in its reaction with copper metal.

....

9 Why is one metal more reactive than another in this type of reaction?

....

....

Evaluation

1 Explain why it is important that the metals have been cleaned.

....

....

2 Why have the metals potassium and lithium not been used in the experiment?

....

3 Explain why the metal is pushed under the surface of the solution.

....

....

4 Why is 5 minutes allowed before the results are recorded?

....

....

5 Did the results of your experiment match the order of reactivity of metals shown in your chemistry textbook?

....

6 If one or two of the reactions gave a 'wrong' result can you suggest a reason for this?

....

7 Suggest **two** ways the method could have been improved to give results that follow the accepted order of reactivity of the metals.

....

....

GOING FURTHER

Displacement reactions also occur between more reactive halogen molecules and less reactive halogens in a salt (see Experiment 9.2). Does the same chemical change happen to the most reactive halogen as occurred to the most reactive metal that you have investigated? Explain your answer.

....

....

....

....

10.3 What causes rusting?

After a period of time, objects made of iron or steel become coated with rust. The rusting of iron is a serious problem and wastes enormous amounts of money each year. Estimates are difficult to make, but it is thought that more than $2.5 trillion a year is spent worldwide on replacing iron and steel structures.

Rust is an orange–red powder consisting mainly of hydrated iron(III) oxide ($Fe_2O_3.xH_2O$). You will investigate the conditions needed for rusting to occur – this will show ways of *preventing* iron from rusting.

You will carefully control variables so that a valid comparison of different factors can be made.

Rusting is actually speeded up in the presence of salt solution.

KEY TERMS 

rust
reactivity series of metals
sacrificial protection

redox reaction

This helps the movement of electrons in the rusting process, which is a redox reaction.

Rusting of iron can be prevented by putting the iron in contact with a more reactive metal such as zinc.

Aim

To plan an experiment which can be used to investigate the conditions needed for rusting to occur.

SAFETY GUIDANCE

- Eye protection must be worn.
- Cooking oil is flammable and should be kept away from the Bunsen flame.
- Take care with boiling water.
- Anhydrous calcium sulfate – irritant
- Solid sodium chloride – low hazard

Apparatus and chemicals

- Eye protection
- 5 × boiling tubes
- Boiling-tube rack
- Bungs for the boiling tubes
- Tripod
- Gauze
- Bunsen burner
- 5 × shiny iron nails
- Emery paper
- Cooking oil
- Anhydrous calcium sulfate (drying agent)
- Distilled water
- Solid sodium chloride
- Thin strips of zinc metal

TIP

To investigate parts **a** to **c** you need to control some factors. Both water and oxygen are present in air, and oxygen is present in water. It is possible to remove oxygen from water by boiling water for a few minutes. Anhydrous calcium sulfate absorbs water vapour from the air.

How could you prevent oxygen diffusing back into water?

Method 1 – planning

1 Using the information given in the theory section and your knowledge from your chemistry course about rusting, plan a method to investigate whether:
   a both water and oxygen are needed for rusting to occur
   b without oxygen rusting does not occur
   c without water rusting does not occur
   d rusting speeds up in the presence of salt solution
   e a more reactive metal attached to the iron nail will slow down rusting.

....

....

....

....

....

....

....

....

....

....

....

....

2 Predict, with a reason, which of the five situations in points a-e would cause the shiny iron nails to rust.

....

....

3 When you have written down your method, check it with your teacher before you carry it out.

Method 2 – carrying out your plan

Throughout the practical the student should wear eye protection.

1 Using your method (which has been checked by your teacher) set up the five boiling tubes to investigate points a-e.

2 You will need to leave the finished tubes a few days for the rust to start to form before they can be checked and your observations entered into the table in the Observations section.

Observations

Write your observations in the table a week after you set up the tubes.

Tube		Observation
a	water and oxygen	
b	without oxygen	
c	without water	
d	with salt solution	
e	with zinc	

Conclusions

1 In which tube(s) did rusting occur?

....

2 Why did rusting occur in your experiment?

....

....

3 In which tube(s) did rusting *not* occur?

....

4 a Explain your answer to question 3.

....

....

b Include ideas about sacrificial protection.

....

....

....

....

....

....

5 In which tube did most rusting occur?

....

6 Rusting is a redox process. When iron rusts, the iron changes into iron(III) ions to form iron(III) oxide.

a What type of chemical change is this?

....

b Write an ionic half-equation to show this change.

....

c Suggest why rusting is speeded up in the presence of salt solution.

....

....

7 What type of chemical change does the oxygen undergo to form oxide ions, also present in the iron(III) oxide? Write an ionic half-equation to show this change.

....

....

Evaluation

1 How have you ensured that in one tube no oxygen gas will be present? How does your method remove the oxygen gas?

....

....

....

2 How have you ensured that one of the tubes has no water present? How does your method remove the water from the tube?

....

....

....

3 Did you make any changes to your initial procedure as you carried out the practical itself? If so, state what you did differently and why.

....

....

GOING FURTHER

Suggest (with a reason) two other substances that could be used for sacrificial protection of iron and steel structures (such as the hull of a ship).

Suggest why these other substances are not commonly used for this purpose.

....

....

....

11 Chemistry of the environment

11.1 Making a fertiliser

KEY TERMS 

neutralisation
artificial fertiliser
crystallisation
titration
indicator

The use of artificial fertilisers has increased over the years. This is because there is an ever-increasing world population that has to be fed and so an increase in crop production is needed.

Crops remove nutrients from the soil as they grow; these include nitrogen, phosphorus and potassium. Artificial fertilisers are added to the soil to replace these nutrients.

Ammonium sulfate is a widely used nitrogenous fertiliser. It is manufactured by the following neutralisation reaction:

ammonia + sulfuric acid $\rightarrow$ ammonium sulfate

$2NH_3(aq) + H_2SO_4(aq) \rightarrow (NH_4)_2SO_4(aq)$

You are going to make a sample of this fertiliser.

Aim

To make a sample of a nitrogenous fertiliser, by reacting an acid with an alkali by titration.

Apparatus and chemicals

- Eye protection
- Burette
- Burette stand
- 2 × 25 cm^3 measuring cylinders
- 250 cm^3 conical flasks
- White tile
- Filter funnel
- Glass rod
- Evaporating basin
- Bunsen burner
- Tripod
- Gauze
- Heat-resistant mat
- 2 × 400 cm^3 beakers
- 0.05 mol/dm^3 sulfuric acid
- 1 mol/dm^3 aqueous ammonia
- Thymolphthalein indicator

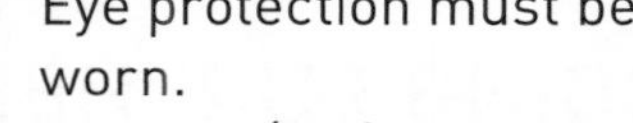

- Eye protection must be worn.
- 0.05 mol/dm^3 sulfuric acid – low hazard.
- 1 mol/dm^3 aqueous ammonia solution – irritant.
- Note: aqueous ammonia releases ammonia – toxic, corrosive (so use in a fume cupboard).
- Thymolphthalein indicator – low hazard.

Method

1 Put on your eye protection.

2 Pour dilute aqueous ammonia into a 25 cm^3 measuring cylinder and pour this exact amount into a conical flask on a white tile to which a few drops of thymolphthalein indicator have been added. Thymolphthalein is blue in alkaline conditions but colourless in acid.

3 Clamp a burette vertically as shown in Figure 1. Use a filter funnel to fill it with 0.05 mol/dm^3 solution of sulfuric acid exactly to the zero mark.

4 Remove the filter funnel.

5 Slowly add the acid from the burette, in small quantities – usually no more than 0.5 cm^3 at a time (Figure 2). Swirl the contents of the flask after each addition of acid to ensure thorough mixing.

6 Add the acid until the aqueous ammonia has been neutralised completely. This is shown by the blue colour of the indicator *just* disappearing.

7 Take the final reading on the burette at the end-point (just as neutralisation takes place). Record this value in the Observation section.

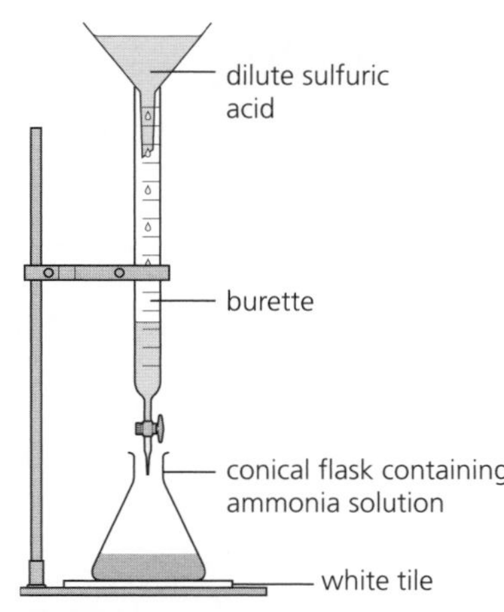

Figure 1

8 Follow these steps to obtain crystals of ammonium sulfate without the thymolphthalein present.
   a To 25 cm^3 of aqueous ammonia in a 400 cm^3 beaker add, while stirring, the volume of sulfuric acid you needed in the titration for neutralisation to take place.
   b Transfer half of this solution to an evaporating basin.
   c Half fill a 400 cm^3 beaker with water and stand it on a tripod and gauze.
   d Place the evaporating basin on top of this beaker as shown in Figure 3. Light the Bunsen burner and evaporate the solution slowly until there is only a small amount left. Do not allow to boil dry.
   e Set aside and leave to crystallise.

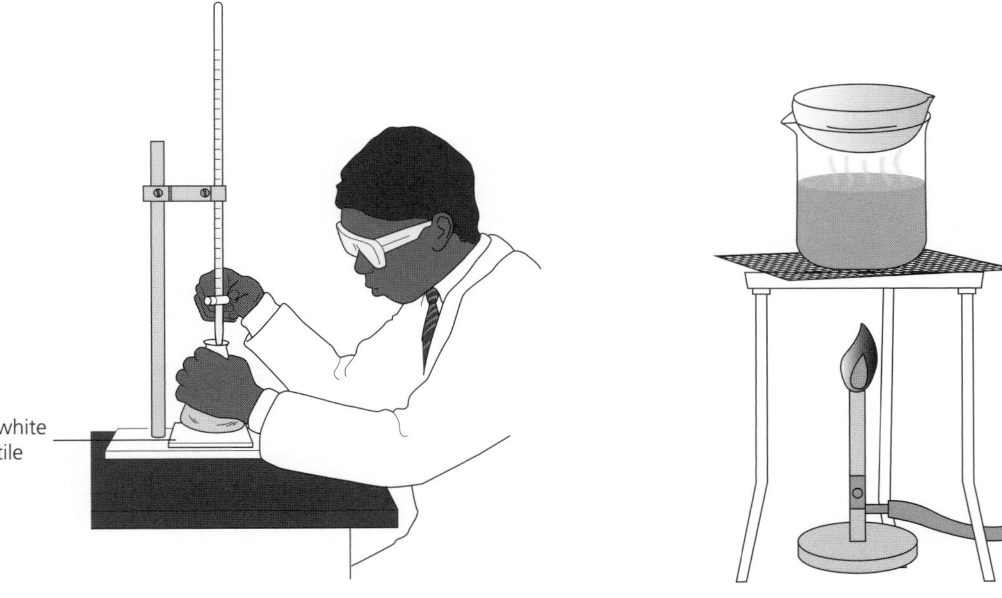

Figure 2

Figure 3

Observation

Volume of dilute sulfuric acid needed for the complete neutralisation of the ammonia solution = ... cm^3.

Conclusion

Crystals of the .... ammonium sulfate can be made by the .... of aqueous ammonia by dilute .... acid.

Evaluation

1 What was the purpose of the white tile?

....

2 Why should the filter funnel be removed from the burette before the titration is carried out?

....

....

3 To make ammonium sulfate crystals, why was the titration process repeated without the indicator present?

....

....

4 Evaluate the method you used and make a list of sources of measurement errors in this experiment.

State at least two ways in which this experiment could have been improved to make sure the volume of acid needed is accurate and state why they would have improved the experiment.

....

....

....

....

....

....

....

....

5 How could the experiment be modified to find out the exact concentration of the aqueous ammonia using 0.05 mol/dm^3 sulfuric acid?

....

....

....

....

....

....

....

....

....

....

6 Describe how you would calculate the percentage yield of the ammonium sulfate formed in this experiment.

Explain why the percentage yield is less than 100%.

....

....

....

....

GOING FURTHER

Ammonium nitrate is another important fertiliser. A student was told to prepare a sample of ammonium nitrate by titrating ammonia solution with dilute nitric acid using methyl orange indicator.

Find out and describe why methyl orange is a suitable indicator for finding the volume of nitric acid that is needed to neutralise a certain volume of ammonia solution.

....

....

....

....

....

12 Organic chemistry 1

12.1 Is the gas burned in Bunsen burners a hydrocarbon? (Teacher demonstration)

A fuel is a substance that can be conveniently used as a source of energy. Fossil fuels release energy in the form of heat when they undergo combustion:

fossil fuel + oxygen → carbon dioxide + water (+ energy)

For example, natural gas is mostly methane (a gaseous alkane). When methane burns in a good supply of air (complete combustion) it forms carbon dioxide and water, as well as plenty of heat energy:

methane + oxygen → carbon dioxide + water (+ energy)

$CH_4(g) + 2O_2(g) \rightarrow CO_2(g) + 2H_2O(g)$

How can we use this reaction to show that methane or the gas used in your Bunsen burner is a hydrocarbon? Your teacher will demonstrate how this can be done.

KEY TERMS 

fossil fuel
alkane
hydrocarbon

Aim

To show that alkanes, such as methane, are hydrocarbons.

Apparatus and chemicals

- Eye protection
- Water pump
- Glass filter funnel
- Safety screen
- 3 × retort stands with clamps and bosses to hold filter funnel and U-tubes
- 400 cm^3 beaker to act as an ice bath
- Ice
- Bunsen burner
- Limewater
- Anhydrous copper(II) sulfate powder

SAFETY GUIDANCE 

- Eye protection must be worn.
- This experiment should be carried out behind a safety screen with the teacher/demonstrator wearing eye protection. Care is needed with the glass funnel. It may crack.
- Limewater – irritant
- Anhydrous copper(II) sulfate powder – corrosive, irritant, dangerous for the environment.

Method

1 The teacher or demonstrator should wear eye protection throughout the demonstration.

2 Set up the apparatus shown in Figure 1 behind a safety screen.

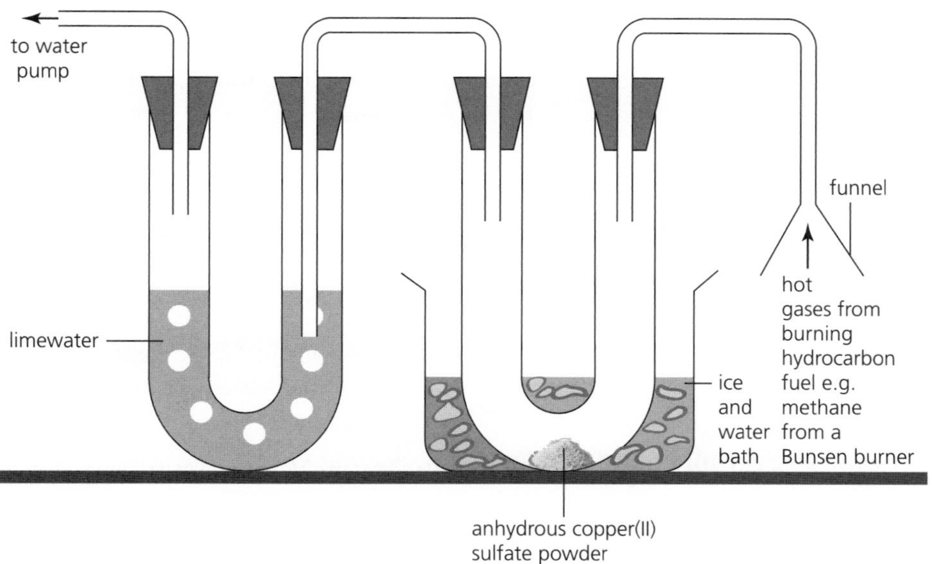

Figure 1

3 Record your observations in the Observations section as the demonstration proceeds.

4 Place a small amount of white anhydrous copper(II) sulfate powder into the bottom of the first U-tube and some limewater in the second U-tube.

Why is the anhydrous copper(II) sulfate powder placed in the first U-tube and the limewater put in the second U-tube?

....

5 Light the Bunsen burner. Set the flame with the air hole half open.

6 Turn on the water pump and then move the Bunsen burner under the filter funnel. Ensure the flame is not too close.

Why is the Bunsen flame not too close to the filter funnel?

....

7 Allow the process to proceed for 5 minutes then turn the Bunsen burner off.

Observations

When the gases produced by the Bunsen burning the fuel in air are passed through the apparatus:

1 What happens to the anhydrous copper(II) sulfate?

....

2 What happens to the limewater?

....

....

Conclusions

Using the observations from this experiment you should now be able to complete the paragraph below as a conclusion to the experiment.

Hint: What does the limewater and anhydrous copper(II) sulfate test tell you about what elements might be present in the burning hydrocarbon fuel?

1 The anhydrous copper(II) sulfate turns from a .... colour to a .... colour.

This shows that .... is present.

When fuels burn they react with oxygen in the air, so this test shows the presence of the element

.... in the fuel.

2 The limewater goes .... The limewater tests for ....

When fuels burn they react with oxygen in the air, so this test shows the presence of the element

.... in the fuel.

The fuel is therefore a ....

Evaluation

1 Outline how the procedure could be improved to make it more reliable.

....

....

2 Methane can also undergo incomplete combustion.

a What is the difference between complete and incomplete combustion?

....

....

b Write a word and a balanced symbol equation for the *incomplete* combustion of methane.

....

....

GOING FURTHER

Natural gas is not pure methane. It contains other hydrocarbons, and often a small amount of sulfur. How could the presence of sulfur in a fuel be detected?

....

....

....

13 Organic chemistry 2

13.1 Reactions of ethanoic acid

Ethanoic acid is an organic acid. It is a member of the homologous series of carboxylic acids with the general formula $C_nH_{2n+1}COOH$. It has typical properties of an acid in that it will react with and be neutralised by alkalis, metal oxides, most metals and carbonates.

KEY TERMS 

carboxylic acid
homologous series
ester

Being an organic acid, it also reacts with some organic molecules such as ethanol, C_2H_5OH. This reaction forms a substance called ethyl ethanoate, which belongs to the homologous series of esters.

You are going to examine some of these reactions in this experiment.
Your teacher will demonstrate the reaction involving the ester formation.

Aim

To examine the reactions of dilute ethanoic acid.

Apparatus and chemicals

- Eye protection
- Test-tube rack
- 4 test tubes
- Test-tube holder
- Thermometer (–10 to 110 °C)
- Delivery tube
- Teat pipette
- 0.4 mol/dm^3 ethanoic acid
- Wooden splints
- Magnesium ribbon
- Sodium carbonate
- Dilute sodium hydroxide
- Ethanol
- 4 mol/dm^3 sulfuric acid
- Limewater
- Access to a kettle for hot water for the water bath

SAFETY GUIDANCE 

- Eye protection must be worn.
- Limewater – irritant
- Sodium carbonate solution – low hazard
- 0.4 mol/dm^3 ethanoic acid – irritant
- Magnesium ribbon – low hazard
- 0.4 mol/dm^3 sodium hydroxide – corrosive
- Ethanol – highly flammable
- 4 mol/dm^3 sulfuric acid – corrosive. Teacher demonstration only.

Method 1 – student practical

1 Put on your eye protection. As you work through the experiments, record your observations.

2 Place a piece of magnesium ribbon into a test tube and add a small amount of ethanoic acid to it. Put your thumb over the end to stop any gas escaping and test the collected gas with a lighted splint.

3 Place a small sample of the sodium hydroxide solution into a test tube in the test-tube rack. Insert the thermometer and record the temperature. Add a small amount of ethanoic acid to it. Measure and record the temperature as the reaction occurs.

4 Place a little sodium carbonate into a test tube and add a small amount of ethanoic acid to it. Put the delivery tube into the neck and pass any gas through limewater.

Why is a delivery tube needed in this reaction?

....

Method 2 – demonstration, ester production

1 Place a little ethanol into a test tube and add a small amount of ethanoic acid, followed by a few drops of 4 mol/dm^3 sulfuric acid, using a teat pipette.

2 After a couple of minutes, pour the mixture carefully into 0.5 mol/dm^3 sodium carbonate solution in a beaker. Then cautiously smell the contents.

Observations

Substance	Reaction with dilute ethanoic acid	Conclusion
Magnesium		
Sodium hydroxide		
Sodium carbonate		
Ethanol		

Conclusion

1 Complete the sentences.

Ethanoic acid is a .... acid. It reacts with magnesium to produce a .... and releases .... gas.

When it reacts with sodium hydroxide, the .... reaction forms sodium .... and water only. In the reaction with sodium carbonate, it forms sodium .... and releases the gas ....

In the final experiment with ethanol, ethyl ethanoate, an ...., is produced.

2 Write the word equations for the reactions that take place in the four experiments.

a With magnesium:

....

b With sodium hydroxide:

....

c With sodium carbonate:

....

d With ethanol:

....

3 Write the symbol equations, with state symbols, for the reactions that take place in the four experiments.

a With magnesium:

....

b With sodium hydroxide, NaOH(aq):

....

c With sodium carbonate, Na_2CO_3(aq):

....

d With ethanol:

....

4 Use your textbook to help you write ionic half-equations for these reactions.

a With magnesium:

....

b With sodium hydroxide:

....

c With sodium carbonate:

....

Evaluation

1 How could you tell in step 3 of Method 1 that a chemical reaction had taken place?

....

2 Why is it necessary to add 4 mol/dm³ sulfuric acid in step 1 of Method 2?

....

3 Propanoic acid, C_2H_5COOH, is another member of the carboxylic acids. Predict the names of the products formed in the reaction of propanoic acid with:

a magnesium.

....

b sodium hydroxide.

....

c sodium carbonate.

....

4 State, with reasons, how you would expect the observations to be different for the reactions of hydrochloric acid with magnesium, with sodium hydroxide and with sodium carbonate.

....

....

14 Experimental techniques and chemical analysis

14.1 Extracting sodium chloride from rock salt

Sodium chloride is an important raw material and could be produced by neutralisation (Experiment 8.1, Determination of the concentration of a solution of hydrochloric acid). However, rock salt contains sodium chloride and is easily mined from the ground, and because it is a soluble salt, it can be separated from the contaminating earthy materials, which are insoluble. This makes the method cheap and simple. Sodium chloride crystals may then be produced.

KEY TERMS

soluble
filtration
filtrate
crystallisation
pure substance
mixture

Aim

To purify rock salt to obtain sodium chloride crystals.

Apparatus and chemicals

- Eye protection
- Retort stand, clamp and boss
- Pestle and mortar
- Tripod
- Gauze
- Glass rod
- Evaporating basin
- 100 cm^3 beaker
- 2 × 400 cm^3 beakers
- Bunsen burner
- Heat-resistant mat
- Rock salt
- Filter funnel
- Filter papers

SAFETY GUIDANCE 

- Eye protection must be worn.
- Rock salt – low hazard
- Take care when handling the hot glassware and hot liquids. Avoid direct contact with the beaker or evaporating basin until it is cooled, to avoid burns.
- You should stand while carrying out the experiment so that you can move away instantly if the beaker or evaporating basin containing the hot liquid breaks or is knocked over.
- Do not get too close when heating to evaporate the filtrate, as the salt solution may 'spit' solution or salt particles.

Method

1 Put on your eye protection.

2 If you are provided with large, rock-like samples then you will have to crush these before you start, using a pestle and mortar (Figure 1).

3 Place some of the rock salt powder you have produced into a 100 cm^3 beaker and half fill with water.

4 Stand the beaker on a tripod and gauze and warm it gently with a Bunsen burner, stirring all the time with a glass rod.

5 After approximately 15 minutes, turn off the Bunsen burner and leave the beaker to stand until it has cooled down, and the insoluble sandy/earthy material has had a chance to settle.

6. Set up the filter funnel and filter paper as shown in Figure 2 and filter the cooled solution from step 5 into a 400 cm^3 beaker.
7. Pour some of the filtrate into an evaporating basin. Half fill a 400 cm^3 beaker with water and stand it on a tripod and gauze.
8. Place the evaporating basin on top of this beaker as shown in Figure 3, light the Bunsen burner and heat to evaporate the filtrate slowly until there is only a small amount of solution left.
9. Set aside and leave to crystallise.

Figure 1 Using a pestle and mortar

Figure 2 Filtration

Observations

Record your observations.

Figure 3 Crystallisation

Conclusion

Can you explain what you have observed?

Evaluation

1 Why was water added to the ground-up mixture?

....

2 Why was the solution put to one side in step 9, rather than being evaporated to dryness?

....

....

3 Give at least one reason why more salt could have been extracted from the original sample of rock salt.

....

....

4 Give at least one reason why the sample of salt you have obtained might not be pure sodium chloride.

....

....

5 Use your answers to write down at least two ways in which this experiment could have been improved and state why they would have improved the experiment.

....

....

....

GOING FURTHER

1 You have been asked to extract sugar from a sugar/sand mixture. Explain how you would adapt the procedure you used for rock salt, and the reasons for differences, if any, there would be in the procedure.

....

....

....

....

....

....

2 The experiment you have carried out would be suitable to obtain very small quantities of sodium chloride. However, companies need to extract very large quantities of sodium chloride. How could you do this on a large scale?

....

....

....

....

....

....

14.2 Chromatography

KEY TERMS

chromatography
mixture
R_f *value*

You may have two or more solids, which are soluble, to separate. This situation arises, for example, when you have mixtures of coloured materials such as dyes. There are also mixtures of substances, such as proteins, which are colourless. The technique that is widely used to separate these materials so they can be identified is chromatography.

In this experiment you will be using ascending chromatography. As the solvent moves up the paper, the dyes, for example, in a mixture are carried with it and begin to separate. They separate because the substances have different solubilities in the solvent and are absorbed to different degrees by the chromatography paper. As a result, they are separated gradually as the solvent moves up the paper.

The chromatogram in Figure 1 shows how a mixture of coloured substances in black felt-tip ink contains three dyes, P, Q and R.

Numerical measurements known as R_f values can be obtained from chromatograms. An R_f value is defined as the ratio of the distance travelled by the solute (for example P, Q or R) to the distance travelled by the solvent. These R_f values are found in data-books for a variety of substances. This is useful for identification purposes.

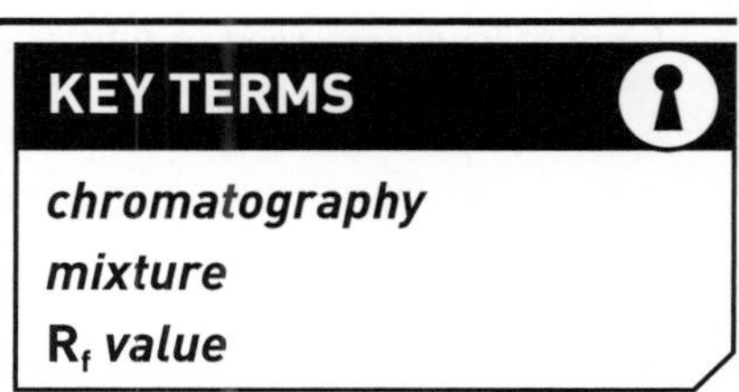

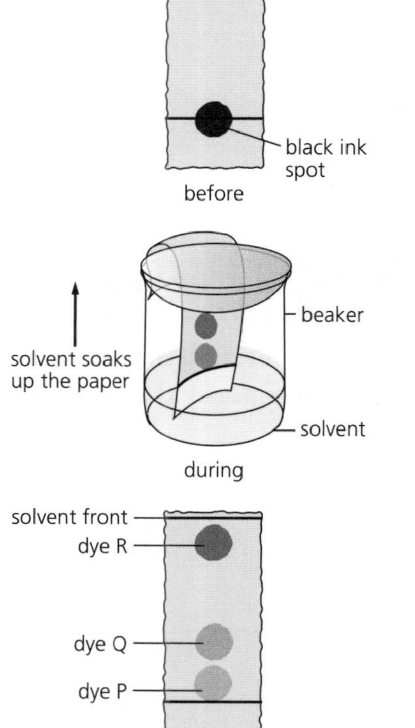

Figure 1

Aim

To investigate the use of chromatography in separating substances in a mixture.

Apparatus and chemicals

- Eye protection
- Piece of chromatography paper (14 × 4 cm)
- 250 cm^3 beaker
- Glass rod
- Two pieces of sticky tape
- Ethanol (or colourless methylated spirits (IMS or IDA))
- Propanone (optional)
- Water
- Plastic bag
- Black felt-tip pen

SAFETY GUIDANCE 

- Eye protection must be worn.
- Propanone – highly flammable, irritant
- Ethanol – highly flammable. Ensure there are no flames present when carrying out this experiment and that the room is well ventilated. Inhalation of solvents should be avoided.

Method

1. Put on your eye protection.
2. Rule a pencil line 1 cm from the bottom of a 14 × 4 cm strip of chromatography paper. This line is called the origin. Place one spot of black felt-tip ink in the middle of this line and leave it to dry.
3. Put 0.5 cm depth of an equal mixture of water and ethanol in the bottom of a 250 cm^3 beaker and hang the chromatography paper as shown in Figure 2. Cover the apparatus with an inverted plastic bag, or a watch glass.
4. When the solvent has risen almost to the glass rod, take out the chromatogram and put it to one side to dry.

5. Remember to mark the solvent front with a pencil as you are going to work out the R_f values later.

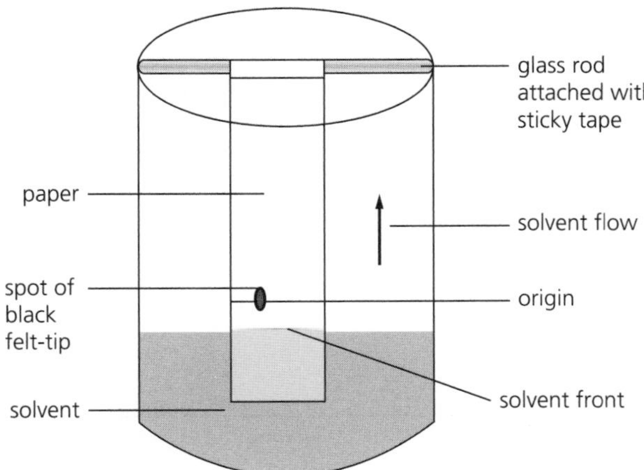

Figure 2

Observations

1 Make a simple sketch of your results, similar to Figure 3.

2 How many spots does the ink separate into?

....

3 Make a note of your observations on Figure 3 below and in the table.

To calculate R_f values you must make measurements of:

- distance solvent front moved.
- distance dyes in mixture travelled – remember to measure to the point where there is the densest colour. If a spot of uniform colour density is formed, measure to the centre of the spot.

In the table, record the R_f values to two decimal places.

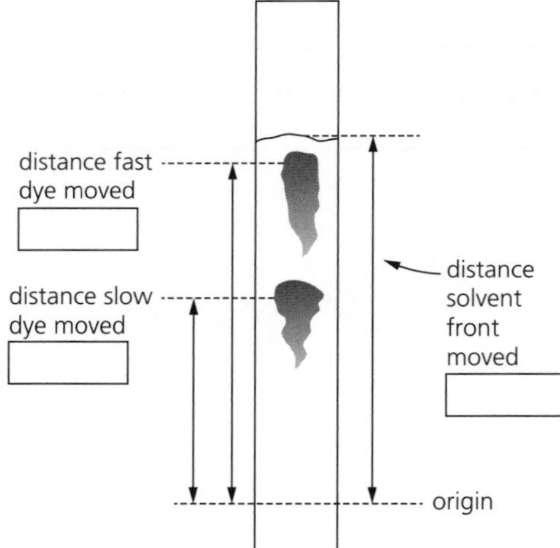

Figure 3

Dye	Distance moved by solvent / mm	Distance moved by dye / mm	R_f value = $\frac{\text{distance travelled by dye}}{\text{distance travelled by solvent}}$

TIP 

Use the same units, mm or cm, for both measurements. Remember that it doesn't matter which units you use, so long as they are both the same!

Conclusions

1 Explain what you have observed.

....

....

....

....

2 Is the dye in the black felt-tip pen a pure substance, or a mixture? How do you know?

....

....

....

3 Which dye was the least soluble and which was the most soluble in the solvent mixture?

....

....

4 Explain why the spots you observed have different R_f values.

....

....

Evaluation

1 Why did you use a pencil line for the origin instead of a ball-point pen?

....

2 Why did you allow the sample spot to dry?

....

3 Why was ethanol added to the water?

....

4 Why was a plastic bag placed over the top of the experimental set-up?

....

5 Why did you allow the chromatogram to dry?

....

6 Describe how you could test black ink from a different pen to find out if they use the same ink.

....

....

....

....

7 Describe how you could use the results to identify a particular substance within a mixture using R_f values.

....

....

....

GOING FURTHER

Predict, with reasons, how the results obtained would be different if the experiment were repeated using a different solvent mixture, such as water, or a mixture of water and propanone.

....

....

....

....

14.3 Missing labels from reagent bottles: what a problem!

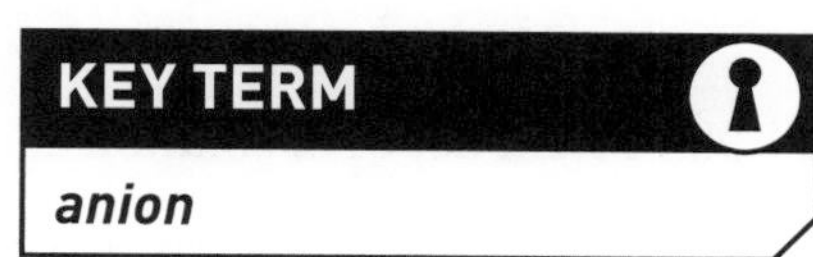

In this experiment, an everyday problem is investigated – a problem has arisen due to missing labels from reagent bottles A, B, C and D! The bottles were known to contain solutions of sodium sulfate, sodium chloride, sodium sulfite and sodium carbonate. However, there were labels for sodium iodide and sodium bromide also present. Can you identify which solution is in which bottle? In this type of analysis, it is only necessary to confirm the presence of certain ions. There are specific chemical tests you can use.

Test for a sulfate: If you take a solution of a suspected sulfate and add it to a solution of a soluble barium salt (such as barium nitrate), then a white precipitate of barium sulfate will be produced.

A few drops of dilute hydrochloric acid are also added to this mixture. If the precipitate does not dissolve, then it is barium sulfate and the unknown salt was, in fact, a sulfate.

Test for a halide: If you take a solution of a suspected halide and add to it a small volume of dilute nitric acid, to make an aqueous acidic solution, followed by a small amount of a solution of a soluble silver salt (such as silver nitrate), a white precipitate of silver chloride will be produced if the halide present is a chloride. However, if a bromide is present then the precipitate seen will be cream in colour. If an iodide is present then the precipitate seen will be yellow.

Test for a carbonate: If a small amount of an acid is added to some of the suspected carbonate (either solid or in solution), then effervescence occurs. If it is a carbonate, carbon dioxide gas is produced, which will turn limewater 'milky' (a cloudy white precipitate of calcium carbonate forms).

Test for a sulfite: If a small amount of dilute hydrochloric acid is added to some of the suspected sulfite (either solid or solution) and warmed gently then effervescence occurs. If it is a sulfite then sulfur dioxide is produced. This gas will turn acidified aqueous potassium manganate(VII) from purple to colourless.

Aim

To use qualitative analysis to identify unknown solutions.

Apparatus and chemicals

- Eye protection
- 4 × test tubes
- Test-tube rack
- Dropping pipettes
- 10 cm^3 of 0.15 mol/dm^3 solutions A, B C and D
- 0.4 mol/dm^3 nitric acid
- 0.1 mol/dm^3 barium nitrate solution
- 0.1 mol/dm^3 silver nitrate solution
- 1 mol/dm^3 hydrochloric acid
- Distilled/deionised water
- 0.1 mol/dm^3 acidified potassium manganate(VII) solution
- Limewater

SAFETY GUIDANCE 

- Eye protection must be worn.
- Nitric acid (0.4 mol/dm^3) – irritant
- Hydrochloric acid (1 mol/dm^3) – irritant
- Limewater – irritant
- Acidified potassium manganate(VII) solution – low hazard
- Barium nitrate solution – low hazard
- Silver nitrate solution – irritant
- Solution A – low hazard
- Solutions B, C – low hazard
- Solution D – harmful, irritant

Method

Test a

1 Put on your eye protection.

2 Place 1 cm depth of solution A in a test tube.

3 To the solution, add a few drops of dilute nitric acid followed by a few drops of silver nitrate solution.

4 Record your observations in the table provided.

What are you testing for with Test a?

....

Test b

1 Put on your eye protection.

2 Place 1 cm depth of solution A in a test tube.

3 To the solution, add about 1 cm^3 of dilute hydrochloric acid.

4 If a gas is produced pour it into limewater solution and shake.

5 Record your observations in the table provided.

What are you testing for with Test b?

....

Test c

1 Put on your eye protection.

2 Place 1 cm depth of solution A in a test tube.

3 To the solution, add a few drops of barium nitrate solution.

4 Now add a few drops of dilute hydrochloric acid.

5 Record your observations in the table provided.

What are you testing for with Test c?

....

Test d

1 Put on your eye protection.

2 Place 1 cm depth of solution A in a test tube.

3 To the solution, add a few drops of dilute hydrochloric acid solution.

4 Now add a few drops of acidified potassium manganate(VII) solution

5 Record your observations in the table provided.

What are you testing for with Test d?

....

Wash out your test tubes with distilled/deionised water and then repeat Tests a, b, c and d with solutions B, C and D.

Observations

Unknown solution	Test	Observation	Conclusion
A	a		
	b		
	c		
	d		
B	a		
	b		
	c		
	d		
C	a		
	b		
	c		
	d		
D	a		
	b		
	c		
	d		

Conclusions

1 Identify the solutions.

Solution A is ____________________.

Solution B is ____________________.

Solution C is ____________________.

Solution D is ____________________

2 Write ionic half-equations for the results you have obtained for A, B, C and D.

....

....

....

....

....

....

Evaluation

1 Explain why you may not need to carry out all the tests on an unknown solution to identify it.

....

....

....

....

2 Give a reason why it may be difficult to tell the difference between tests for very dilute solutions of chloride, bromide and iodide ions.

....

....

....

3 Describe **two** procedures that could be used to identify these substances, and (possibly) suggest advantages and disadvantages of each.

....

....

....

....

....

....

14.4 Using flame colours to identify unknown metal ions

If a clean nichrome or platinum wire is dipped into a metal compound and then held in the hot part of a Bunsen flame, the flame can become coloured. For example, a compound of lithium gives a red colour in a Bunsen flame. Certain metal ions may be detected in their compounds by observing their flame colours. Forensic scientists use this technique on an everyday basis to identify the metal ions in substances left at crime scenes.

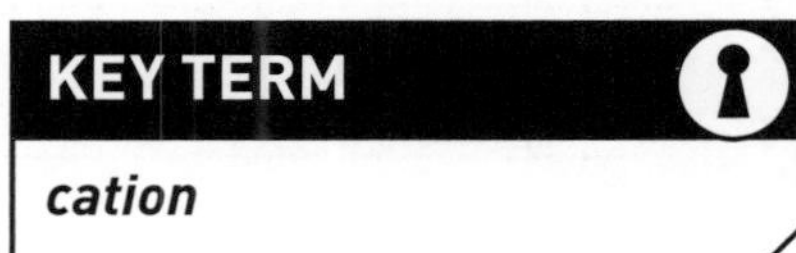

You are given six solutions. Each one has a different metal ion present in the compound found in that solution. Using flame tests you are expected to identify the metal ions present.

Aim

To use flame tests to identify metal ions in solution.

Apparatus and chemicals

- Eye protection
- Bunsen burner
- Heat-proof mat
- Watch glass
- Test-tube rack
- Test tube containing dilute hydrochloric acid
- Flame test wire in holder (An alternative to using flame test wires is to use wooden splints soaked in distilled water overnight.)
- Access to six solutions to investigate, labelled A, B, C, D, E, F.

SAFETY GUIDANCE 

- Eye protection must be worn.
- Hydrochloric acid (1 mol/dm^3) – irritant
- Some of the unknown solutions are low hazard but some are multi-hazard, so treat all solutions as potentially having many different hazards including irritant, corrosive and harmful.
- Hot wire can cause serious burns. Allow equipment to cool on a heat-proof mat before touching.

Method

1. Put on your eye protection.
2. You have six solutions labelled A, B, C, D, E, F.
3. Collect a sample of solution A.
4. Clean the flame test wire. This is done by placing it in a roaring Bunsen flame (just above the blue cone, as shown in Figure 1) and then into dilute hydrochloric acid in a watch glass. The wire should then be put back into the Bunsen flame to ensure that it is completely free from contaminating substances.
5. Dip the clean flame test wire into solution A.
6. Put the flame test wire into the Bunsen flame. Record the colour of the flame in the Observations and conclusions table provided The colour of the flame gives you an indication of the metal ion present in the solution.
7. Repeat steps 3–6 with samples B, C, D, E, F. Record your results in the table. Remember it is important to clean the wire between each test solution and also to use a fresh sample of acid for the cleaning process.
8. Use your textbook to help you identify the metal ions present.

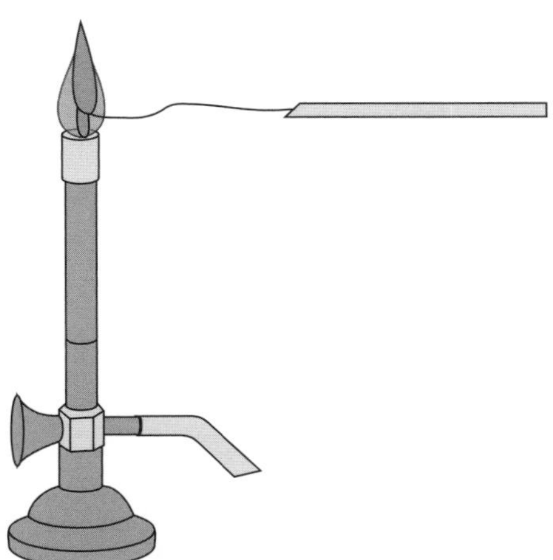

Figure 1

Observations and conclusions

Solution	Flame colour	Metal ion present
A		
B		
C		
D		
E		
F		

Evaluation

1 Why is the wire cleaned in the position just above the blue cone of a Bunsen flame?

....

2 Why do you use a solution rather than solid samples?

....

3 Why must a flame test wire be made of nichrome or platinum, and not copper?

....

....

4 Outline how this experiment could be improved to make it more reliable.

....

....

5 Describe another test you could use to confirm the presence of the metal ions you have identified. Explain why not all the metal ions can be identified using this second test.

....

....

6 Outline some limitations of the flame test method for identifying metal ions in a solution.

....

....

....

14.5 Investigating the solubility of sodium chloride in water

The solubility of a substance is defined as the amount of the substance, the solute, which can dissolve in water, the solvent, at a particular temperature. The temperature must always be recorded as it is usual that most solutes become more soluble at higher temperatures. The solubility of a substance in water is usually expressed as the mass in grams of the substance which will dissolve in 100 g of water (g/100 g H_2O).

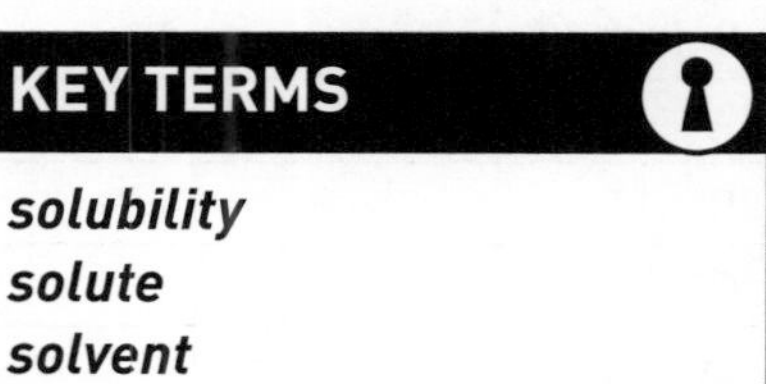

Aim

To find the solubility of sodium chloride at room temperature.

SAFETY GUIDANCE 

- Eye protection must be worn.
- Sodium chloride is not harmful.

Apparatus and chemicals

- Eye protection
- 50 g of sodium chloride in a beaker
- 100 cm^3 beaker
- Spatula
- Top pan balance (minimum resolution 0.1 g)
- Glass rod
- Thermometer
- Measuring cylinder – 50 or 100 cm^3

TIP 

You are going to find the mass of solid that will dissolve in a certain mass of water, so it is essential to measure the volume of water accurately. Use a measuring cylinder, not the markings on the beaker.

Method

Throughout the practical the student should wear eye protection.

1 Use the top pan balance to measure the initial mass of the sodium chloride and its beaker. Record this mass in the Observations section.

2 Place 50 cm^3 of water into the 100 cm^3 beaker. Place the glass rod in the water.

3 Use the thermometer to measure the temperature of the water. Record the temperature in the Observations section.

4 Add a spatula of sodium chloride to the water. Stir the solution with the glass rod until the sodium chloride has fully dissolved.

5 Continue to add sodium chloride to the water, stirring all the time until you can no longer get the sodium chloride to dissolve and some remains at the bottom of the beaker.

6 Use the top pan balance to record the final mass of the remaining sodium chloride solid and beaker. Record this mass in the Observations section.

Observations

Record your results in the table.

Temperature of the water / °C	
Initial mass of sodium chloride and beaker / g	
Final mass of sodium chloride and beaker / g	

Conclusions

1 Use your results to calculate the mass of sodium chloride that dissolved in the water.

....

2 The density of water is 1.0 g/cm^3. What mass of water was used in the experiment?

....

3 Calculate the solubility of the sodium chloride, using the equation below:

$$\text{solubility of sodium chloride (g/100g } H_2O\text{)} = \frac{\text{mass of sodium chloride dissolved (g)}}{\text{mass of water used (g)}} \times 100$$

Give your result to two significant figures.

Solubility of sodium chloride is: .... g/100g H_2O at .... °C.

Evaluation

1 Look at the value you have obtained for the solubility of sodium chloride and compare it with the values from other students in your class. Can you suggest any reasons for the differences?

....

....

....

....

2 Identify **one** source of possible inaccuracy in the results.

....

3 Describe how you could adapt the method to reduce this source of error and explain why this would be an improvement.

....

....

GOING FURTHER

It was stated earlier that the solubility of a solute increases at higher temperatures. Describe how you could change the method given to allow you to produce a solubility curve, which is a graph showing how the solubility of a solute changes with temperature.

....

....

....

....

Past paper questions

Practical test past paper questions

1 You are going to investigate the reaction between aqueous sodium carbonate and two different solutions of dilute hydrochloric acid labelled **A** and **B**.

Read all the instructions carefully before starting the experiments.

Instructions

You are going to carry out three experiments.

a *Experiment 1*
Use a measuring cylinder to pour 25 cm^3 of aqueous sodium carbonate into a conical flask.
Add ten drops of thymolphthalein indicator to the conical flask.
Fill the burette provided up to the 0.0 cm^3 mark with solution **A** of dilute hydrochloric acid.
Add solution **A** from the burette while swirling the flask, until the solution just changes colour.
Record the burette readings in the table below.
Experiment 2
Empty the conical flask and rinse it with distilled water.
Repeat Experiment 1 using methyl orange indicator instead of thymolphthalein.
Record the burette readings in the table below and complete the table. [4]

	Experiment 1	Experiment 2
final burette reading/cm^3		
initial burette reading/cm^3		
difference/cm^3		

b *Experiment 3*
Empty the conical flask and rinse it with distilled water.
Pour away the contents of the burette and rinse the burette with solution **B** of dilute hydrochloric acid.
Repeat Experiment 1 using solution **B** instead of solution **A**.
Record the burette readings in the table below and complete the table. [2]

	Experiment 3
final burette reading/cm^3	
initial burette reading/cm^3	
difference/cm^3	

c What colour change was observed in the flask in Experiment 1?
from .... to .... [1]

d State **one** observation, other than colour change, when hydrochloric acid was added to sodium carbonate. [1]

....

e Complete the sentence below.
Experiment .... needed the largest volume of hydrochloric acid to change the colour of the indicator. [1]

f What would be a more accurate method of measuring the volume of the aqueous sodium carbonate? [1]

....

g What would be the effect on the results, if any, if the solutions of sodium carbonate were warmed before adding the hydrochloric acid? Give a reason for your answer.

effect on results ....

reason .... [2]

h **i** Determine the ratio of volumes of dilute hydrochloric acid used in Experiments 1 and 3. [1]

....

ii Use your answer to **(h)(i)** to deduce how the concentration of solution **A** differs from that of solution **B**. [1]

....

i Suggest a **different** method, using standard laboratory chemicals, to determine which of the solutions of dilute hydrochloric acid, **A** or **B**, is more concentrated. [3]

....

....

....

....

....

j Hydrochloric acid is hazardous.
Suggest **one** safety precaution to follow when using hydrochloric acid. [1]

....

[Total: 18]

(Cambridge IGCSE Chemistry 0620, Paper 51 Q1 June 2016)

2 A sample of furniture cleaner contains aqueous sodium chloride, aqueous ammonia and sand.

a Give a test to show the presence of ammonia in the mixture. [1]

....

b Plan experiments to obtain a sample of

i pure water from the mixture .... [2]

....

....

....

ii pure sand from the mixture .... [3]

....

....

....

[Total: 6]

(Cambridge IGCSE Chemistry 0620, Paper 51 Q3 June 2017)

3 You are going to investigate what happens when dilute hydrochloric acid and copper(II) sulfate solution react with different metals.

Read all the instructions carefully before starting the experiments.

Instructions

You are going to carry out five experiments.

a *Experiment 1*

Use a measuring cylinder to pour 10 cm^3 of dilute hydrochloric acid into a boiling tube. Put the boiling tube into a rack for support.

Measure the temperature of the hydrochloric acid and record it in the table below.

Add 1 g of zinc to the boiling tube and stir the mixture with the thermometer.

Measure and record the maximum temperature reached by the mixture. Pour the mixture away and rinse the boiling tube.

Experiment 2

Repeat Experiment 1 using 1 g of iron instead of zinc.

Record your results in the table.

Experiment 3

Repeat Experiment 1 using 1 g of magnesium instead of zinc.

Record your results in the table. Complete the final column in the table.

experiment	initial temperature of acid/°C	maximum temperature reached/°C	temperature rise/°C
1			
2			
3			

[3]

b *Experiment 4*

Use a measuring cylinder to pour 10cm^3 of copper(II) sulfate solution into a boiling tube.

Measure the temperature of the solution and record it in the table on page 140.

Add 1 g of magnesium to the boiling tube and stir the mixture with the thermometer.

Test the gas given off with a splint and record your result in the space below. Measure the maximum temperature reached by the mixture and record it in the table. Pour the mixture away and rinse the boiling tube.

test ....

result .... [1]

c *Experiment 5*

Repeat Experiment 4 using 1 g of iron instead of magnesium. You do **not** need to test the gas.
Record your observations in the space below and record your temperatures in the table.
Complete the final column in the table.

experiment	initial temperature/°C	maximum temperature/°C	temperature rise/°C
4			
5			

observation .... [3]

d Draw a labelled bar chart for the results of Experiments 1, 2, 3, 4 and 5 on the grid. [3]
Use your results and observations for Experiments 1, 2 and 3 to answer the following questions.

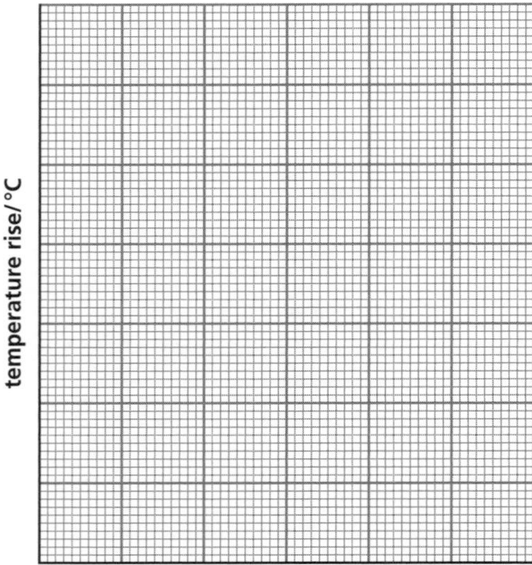

e **i** Which experiment, 1, 2 or 3, produced the largest temperature rise?.. [1]

ii Suggest why this experiment produced the largest temperature rise.. [1]

f Name the gas given off in Experiment 4.

.... [1]

g Suggest why potassium was **not** used as one of the metals in these experiments.

.... [1]

h Give **one** advantage of using a measuring cylinder to add the hydrochloric acid to the boiling tube.

.... [1]

i Suggest and explain **one** change to improve the accuracy of these experiments. [2]

....

....

....

[Total: 17]

(Cambridge IGCSE Chemistry 0620, Paper 53 Q1 June 2016)

Alternative to Practical past paper questions

1 The diagram shows the apparatus used to separate a mixture of water, boiling point 100 °C, and ethanol, boiling point 78 °C.

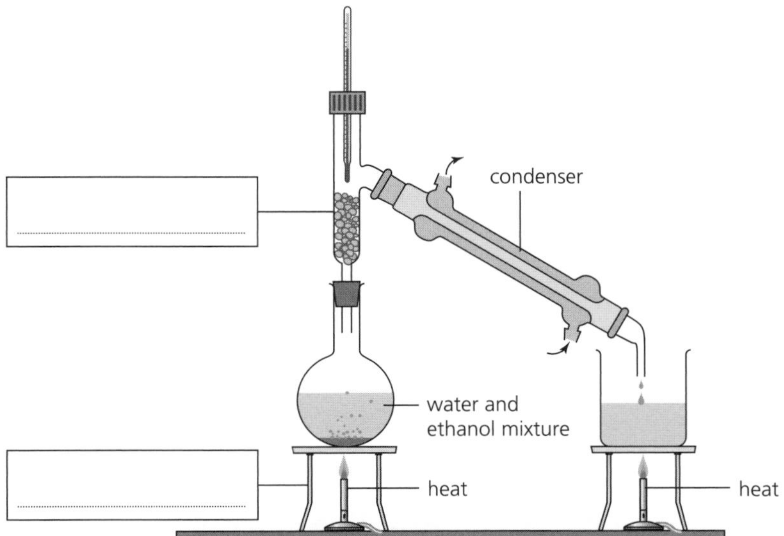

a Complete the boxes to name the apparatus. [2]

b Label the arrows on the condenser. [1]

c Identify **one** mistake in the apparatus. [1]

....

d Which liquid would collect first? Explain your answer. [2]

....

....

e Why would it be better to use an electrical heater instead of a Bunsen burner to heat the water and ethanol mixture? [1]

....

[Total 7]

(Cambridge IGCSE Chemistry 0620, Paper 61 Q1 June 2016)

2 Calcium burns in air to form calcium oxide. The reaction is vigorous and some of the calcium oxide can be lost as smoke.

Plan an investigation to determine the maximum mass of oxygen that combines to form calcium oxide when 2 g of calcium granules are burnt in air.

You are provided with common laboratory apparatus and calcium granules. [6]

....

....

....

....

....

....

....

....

[Total 6]

(Cambridge IGCSE Chemistry 0620, Paper 61 Q4 June 2016)

3 A student prepared strontium nitrate crystals.

The diagram shows some of the stages in this preparation.

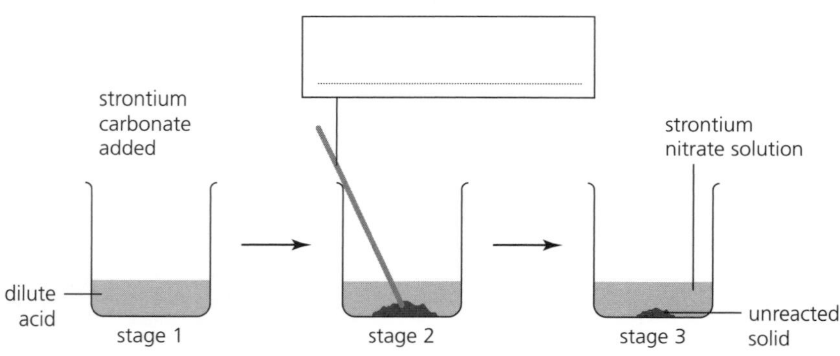

a i Complete the box to identify the apparatus. [1]

ii What is used to add the strontium carbonate to the acid in stage 1? [1]

....

iii Name the dilute acid used. [1]

....

iv Give **one** expected observation in stage 2. [1]

....

b Why is heat **not** necessary in stage 2? [1]

....

c Which of the reactants is in excess? Explain your answer. [2]

....

....

d Describe how crystals of strontium nitrate could be obtained from the mixture in stage 3. [3]

....

....

....

[Total 10]

(Cambridge IGCSE Chemistry 0620, Paper 61 Q1 June 2017)

4 The volume of dilute nitric acid that reacts with 25.0 cm^3 of aqueous potassium hydroxide can be found by titration using the apparatus shown.

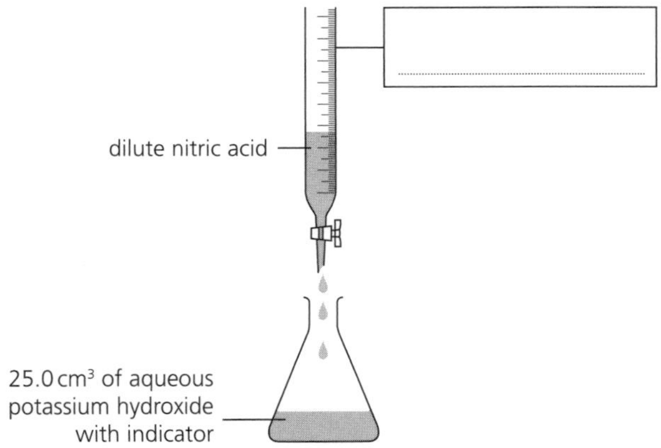

dilute nitric acid

25.0 cm^3 of aqueous potassium hydroxide with indicator

a Complete the box to name the apparatus. [1]

b Name a suitable indicator that could be used. [1]

....

A student did the titration four times and recorded the following results.

titration number	volume of dilute nitric acid / cm^3
1	18.1
2	18.9
3	18.3
4	18.2

c **i** Which **one** of the results is anomalous? [1]

....

ii Suggest what might have caused this result to be anomalous. [1]

....

iii Use the **other** results to calculate the average volume of dilute nitric acid that reacted with the aqueous potassium hydroxide. [2]

....

d The equation for the reaction taking place in the titration is shown.

$HNO_3 + KOH \rightarrow KNO_3 + H_2O$

The student concluded that the aqueous potassium hydroxide was more concentrated than the dilute nitric acid.

Explain whether or not the student's conclusion was correct. [2]

....

....

....

[Total 8]

(Cambridge IGCSE Chemistry 0620, Paper 61 Q1 June 2018)